Rescue
at Engine 32

To Jessica & Sienna (when she is
21 yrs old)

XO X
Jessica Locke

Printed in the United States of America
Third Edition

Quote by F.M. Alexander from
Man's Supreme Inheritance
used by permission of Mouritz, London © 1996

Don't Look For Me Anymore
used by permission of Alicia E. Vasquez © 2001

We Draw Strength From Each Other
used by permission of Robert Girandola © 2006

Book design by Drew Allison
Cover photograph by Jessica Locke
Back cover photograph by Maria Scarfone

A PORTION OF THE PROCEEDS FROM THIS BOOK
WILL BE USED TO ASSIST FDNY FIREFIGHTERS AFFECTED BY 9/11
AND THE JESSICA LOCKE FIREFIGHTERS FUND

ISBN-13: 978-0-9792901-0-7
ISBN-10:   0-9792901-0-4

"God help us if we ever forget 9/11."

–Deputy Chief Roger Sakowich, FDNY

AUTHOR'S NOTE

The Alexander Technique, referred to in this book, is a form of mind/body re-education. It teaches maximally efficient body movement which, in every day activities, results in much less effort and strain.

I like to compare it to the graceful ease of animals. Always light, always free, even after sustaining trauma or injury, animals have a dynamic "primary control" in the way they move which maintains balance against gravity. Human beings have this same control, but may have lost touch with it due to the stresses of modern day life.

Using the body in the way it was designed to function can increase ease of movement, reduce or eliminate chronic pain, improve respiratory function,[1] and reduce or eliminate psychological symptoms such as anxiety, depression and post-traumatic stress disorder.

"Every man, woman and child holds the possibility of physical perfection; it rests with each of us to attain it by personal understanding and effort."

– F. Matthias Alexander

---

[1] In a group study performed by John H.M. Austin, M.D., at the Respiratory Function Laboratory at Columbia-Presbyterian Medical Center, New York, NY, a course of 20 Alexander Technique lessons given to normal healthy adults caused an approximately 10% increase in respiratory muscle function. Control subjects who had no lessons showed no change. The increases in those who had a course of 20 lessons were statistically significant. [Austin JHM, Ausubel P. Enhanced respiratory muscular function in normal adults after lessons in proprioceptive musculoskeletal education without exercises. Chest 1992; 102:486-490.]

PREFACE

There is a sign at Engine 32, located in a stairwell leading to the third floor. It reads:

> WHAT YOU SEE HERE
>
> WHAT YOU SAY HERE
>
> WHAT YOU HEAR HERE
>
> WHEN YOU LEAVE HERE
>
> LET IT STAY HERE

It is a code observed and honored in every firehouse in the City of New York.

That I was permitted within those walls to see what I did, experience what I did, and witness what I did, was precipitated by the disastrous consequences of the terrorist attacks at the World Trade Center on September 11, 2001.

I trust that the firefighters of Engine 32 will forgive that here I break their code, so that ultimately I can deliver to both time and history a true record of the character of those men who gave their lives on that day, and how those who survived conducted themselves in the aftermath of the destruction.

If the firefighters saw it differently than I … well, you will have to ask them.

Only they will never tell you.

Rescue
at Engine 32

Jessica Locke

BOSTON, MASSACHUSETTS
AUGUST 26, 2001

Last night, I had a dream I was dancing with my cat, her front paws in my hands.

My mother enters the room, and with one swift chop of her hand cuts the cat in half. The cat falls to the rug in two pieces. The bottom half of the cat turns black, the upper part turns white. The front half of the cat struggles forward, dragging itself by its claws, as bloody entrails pour out. In horror and grief I look to my mother to do something. *Do something.* The cat must be taken to the vet and sewn back together, or it will die. But she just stands there, a look of sick satisfaction on her face. I cannot speak, nor can I make any move, lest she do the same to me.

GROUND ZERO, NEW YORK CITY
JANUARY 13, 2002

It was a few minutes after twelve on Sunday afternoon. Blue saw-horse barricades blocked all access to Church Street, where two uniformed police officers in navy wool coats stood guard. It was impossible to get a view of the World Trade Center site except from the newly-constructed platform on Fulton Street, and that required a ticket.

I had one for 6:00 p.m.

I arrived from Boston by train the day before. My brother was singing in a chorus at Carnegie Hall on Saturday night. He and his wife were driving back to Ohio after the concert, leaving an empty hotel room already paid for. With my limited finances, it seemed like a good opportunity to be in New York City and complete the research required to compose a sacred orchestral/choral work for September 11th. The only words in the piece would be the names of the nearly 3,000 souls murdered on that day. The working title was *Reading of the Names.*

That was what I was telling everyone. But it wasn't the truth.

Ever since the morning of the attacks on the World Trade Center, I felt like someone had taken a large baseball bat and whacked my heart and soul clean out of my body, sending them to the smoking pile of debris in lower Manhattan. My brain – split in two and frozen emotionless – had only survival instinct remaining. And that instinct heard a "voice" that demanded I go to Ground Zero. If there was any hope of putting myself back together, this voice insisted that I go directly to the site and stand at its center. What kind of insane treasure hunt was this? It called me, pushed me, insisting that I go there. Set a date, make a reservation for the train, and just go.

Problem was, I didn't "just go" anywhere. I rarely drove outside of a 10-mile radius of Boston. My friends would say that I hardly ever left my apartment. It had quietly become automatic to plan my life in avoidance of the possibility of rape or murder, and to minimize the risk. If someone suggested we meet at a bar, I would immediately remind them of the nanny who met someone in a bar, was cut in half and left in a dumpster. You would think I was in charge of security for the President, the way I could review all possibilities of attack in any scenario. The end result was that if there was risk, you could count me out.

So, to go to New York City was insane. The risks were enormous. There could be another terrorist attack. I could be trapped. I could get lost. Someone might steal my train ticket, my money, and I wouldn't be able to get home. I could be raped and murdered, and no one would know what had become of me. I had not made out my will. I had two cats. They would become homeless. And where would I find a bathroom? It was an endless circle of fear and inaction, and the attacks on the World Trade Center had suddenly increased the dynamics to critical mass.

To appease that insistent voice that demanded I *do something*, I fixated on the rescue efforts taking place on my television screen, a feeble attempt to provide moral support while watching these men standing in bucket lines in 12-hour shifts. I stayed glued to the television for weeks, hoping to hear that they had found someone alive; but redemption never came.

The news coverage grew shorter and fainter. "Rescue" became "recovery"; so many people were missing, but where were they? How could they just disappear? I found myself drawn to the tragic loss of the 343 firefighters. Those remaining had made a huge effort to bring honor to the process of finding the dead, while the political pontiffs seemed in too big of a hurry to clean the city of this vile experience, hauling out the unsifted body parts in dumptrucks. There was a stand-off: Eighteen firefighters were arrested. I was sick and disheartened. Nobody needed this. What was the hurry? Why couldn't they let them do this right?

My best friend Dawn suggested that since neither of us was capable of giving blood without fainting, we could do our part by stimulating the economy. In late September she took me shopping for a belated 50th birthday present, a beautiful, blue-flowered Calvin Klein duvet cover that I had seen on "Sex in the City." It cost $250, and it was lovely. We were in Bloomingdales on a Saturday, and we were the only customers in the store.

A film and video composer, I had no desire to write music. The painful divide in my head made me doubt my access to the composer side of me. There were no jobs coming in, anyway. As a musician it was necessary to have a second income, so I also gave lessons in the Alexander Technique. Unfortunately, in times of economic fall-out, bodywork was generally the second thing people gave up; music for videos was the first. I was forced to take in a roommate to make the rent.

I turned on the television every morning with the fear and dread that I might witness another attack. I walked to the grocery store and saw everything explode in flames around me. Airplanes flying overhead appeared black and evil. If I drove into Boston, I couldn't look at the Prudential Building without seeing a plane ramming into it. It brought back the memory of my first date at the Top of the Pru with my ex-husband in 1979. There had been a power failure throughout Boston that night and we had to walk down the 52 flights of stairs to the street. It had taken so long, and my legs had given out. I imagined myself in the Twin Towers, trying to add 25 or 30 more floors.

At some time in November it occurred to me that living in sickening fear, feeling intense personal pain, stuffing down my emotions, waiting for the next 'hit,' was not new to me. This was the ugly grasp of my childhood returning. Oh, I get it now: My mother was a terrorist. Oddly enough, coming to this realization provided some comfort. It made me realize that I already knew how to navigate through this, and I began to reach for the skills I had long ago set aside.

I began to wonder what the future would be like for the people in New York who had been raped with such unspeakable evil for the first time. For the firefighters, police and emergency workers, for the families who would live for years with the seemingly unhealable wounds of hate: they had no idea how hard it was going to be to sort it all out.

From these pieces of clarity I realized for the first time that my buried past had a positive value. I could go down to NYC and offer my services not only as a body worker, but as someone who had already been through it all. I could lend a solid – and experienced – hand. I had personally dealt with something this ugly and evil before. I could show them I made it through, and they could, too. I could *do something*. An intense rage rose up inside, and I knew I had to go to NYC and find those bastards who did this and kill them with my bare hands. I heard myself saying out loud – to bin Laden, to my mother: You will never do this to anyone ever again!!!

Talk, talk, talk.

I was a coward, and I wasn't going anywhere.

The weeks after 9/11 grew empty and meaningless. "Go to Ground Zero," the "voice" argued, impatient, restless, and demanding action far beyond my level of courage.

I resisted. Why should I walk again toward such darkness? What would it accomplish? If mankind had reached its highest evolutionary achievement in the 21st Century by flying planes into buildings, well then, we were all doomed. The voice ignored my arguments. "Go to Ground Zero."

It took the unusual invitation from my brother – and the security his presence would provide – to finally give in and make the train reservation to New York. In that moment, I was enveloped by a golden and loving warmth that reassured me I had done the right thing. It bathed me in an all-encompassing peace I had never before experienced.

That, of course, totally freaked me out. I had read somewhere that when you were about to die you would get a telegram from some angel-like messenger. Just like this. And the message was clear: New York City was going to be the place for my death. As long as I stayed in Boston I could delay it, but after a week of painful reflection I had to face the fact that my life had been nothing but a series of failures; climbing up a glass bowl only to slide back down. I had no family of my own, was not even marginally successful as a composer, lived alone with two cats, and there would be no one to take care of me if and when I grew old.

My sister would take my cats if I died.

On January 12, 2002 I packed my bag and got on the train headed to Ground Zero.

MAUMEE, OHIO
JULY 1943

The details of my mother's love affair with Harold are threadbare, but from what I am able to piece together, he was extremely intelligent and well read. He left romantic poems and notes in her locker at school. He could converse on any subject, and they spent hours and hours talking. He was not handsome, and he didn't want children. That alone was enough to convince me that he was the perfect man for her, and if her parents had not intervened, I would not be here to write this.

Harold was Jewish. It was the early 1940's in midwestern Ohio, and they were not about to let their 18-year-old daughter marry a Jew.

Meanwhile, my father's brother had moved to Maumee with his wife and kids, and my mother became their babysitter on a regular basis. One afternoon my father, Sergeant William M. Locke, then age 23, showed up on their front porch, back home from two years overseas in World War II. He was drop-dead gorgeous, a cross between Montgomery Clift, Gregory Peck, and George Clooney.

My mother's parents were greatly impressed with his pedigree; bloodlines to John Locke, the philosopher, and John Bilington, who came over on the Mayflower. Dad's father was a lawyer, as were several of his twelve siblings. There was an assumption that he would pursue a degree in law as well. He held a bachelor's degree in – of all things – philosophy.

(Curious to understand his interest in this discipline, as a freshman

7

attending Ohio State University I enrolled in an introductory course. The professor was always drawing a picture of a tree, and saying that the tree only existed because we agreed in our minds that the tree was there. After several weeks of this, I raised my hand and said, "There's no money in this, is there." The professor froze. The whole lecture hall broke up laughing. I didn't think it was particularly funny, having grown up as poor as we did. Lesson No. 1: Never get a degree in something that won't pay your bills.)

But I digress.

My grandparents were resolute in their belief that this was the man their daughter should marry, and according to my mother, a lot of pressure was steadily applied. The doomed Harold enlisted in the Army, and was sent overseas. I do not recall the number of letters he sent to my mother, only that Gramma had decided to take matters into her own hands and snag the letters out of the mailbox before my mother saw them. My mother, believing he had stopped caring about her, accepted the dazzling consolation prize waiting down the street, and with her parents' two-faced blessing she married my father in 1945.

I don't know the exact timing of all this, but Dad failed the law school exam and became a mattress salesman for Bemco. He and his new bride moved into a made-over chicken coop; two children were born, and sometime later Mom had a nervous breakdown. It all seemed to stem from the day she came upon a pile of unopened letters from Harold stuffed in the back of Gramma's dresser drawer.

A rage was set free that day, a rage that I suspect must have been fomenting for years. Finding herself trapped on a path she had not chosen, cruelly robbed of free will and choice, and unable to see how she could get out, she decided to get even. Her husband and her children were meaningless by-products of her parents' wishes. This was not of her design, and therefore, she was no longer responsible for any of it. She would make all of us pay. I wasn't even born yet.

MIDTOWN MANHATTAN
JANUARY 13, 2002

I checked out of the hotel at precisely 10:59 a.m. Sunday morning after spending the night in absolute terror. It was all I could do to force myself to leave the hotel, to walk out to the street. Whatever was happening here was real, and it was certainly going to kill me. My imagination was running full-tilt with all the possibilities for the seemingly inevitable demise. I did not want to die pushed in the path of a subway train, which I had heard happened to other people at various times over the years. No subway for me. So, the most likely place for my death would involve a cab driver taking me to a remote warehouse to rob, rape and kill me. I would have to take a cab to the South Street Pier to get my ticket for the viewing platform. Then a cab to Ground Zero. And another back to Penn Station. Somewhere along the way I would die. Every move I made now was life and death, and I was sick to my stomach as I raised my hand to signal the next step closer to it.

A yellow taxi swerved over and stopped. I got in and gave the destination of the South Street Pier. The driver, having no imagination whatsoever, took me directly to the Pier and dropped me off without a word. After I picked up my ticket at the kiosk, the only response I got out of Cab Driver #2 was an incredulous look that I would not walk the seven blocks to the World Trade Center site, but then, he didn't understand that I had no idea where I was. He left me off on Broadway amidst a slurry of people and buildings, pointing in the general direction of where St. Paul's Church was supposed to be.

As I approached, I saw a thick, slow-moving line of people trudging toward the viewing platform. The line extended down and around the block. There was no way I would get through that line at 6:00 and make it back to Penn Station in time for my train. I slowly made my way through the crowd and toward the church, the horrors trapped in the atmosphere pulling down so hard on my heart I could barely stand upright.

I stopped short, suddenly not breathing anymore. The manmade memorial at St. Paul's, which stood adjacent to Ground Zero, could barely register on any scale of human emotion. The fence surrounding three sides of the building was covered from top to bottom: 'Missing' posters, pictures of lost loved ones, poems, banners, notes from people around the world, flowers in various stages of life, hats and t-shirts from police and fire stations. All were interwoven through the grille of the fence. At its base: more flowers, more pictures, candles, and teddy bears overflowing the sidewalk like a garden grown wild. I saw a "God Bless New York" banner from Maumee, Ohio – my mother's hometown.

Walking along the memorial wall past the church down Vesey Street, I tried to prepare myself for the first view of the site. It turned out to be not so much a view as it was a sound; a vibration so heavy with grief that I felt pushed backwards by its force.

There was a non-stop, searing, pulsating hum from the engines of countless trucks, derricks, and generators. The invisible current pushed and pulled and pounded at me from all sides, demanding that I let go of my control and flow with it. I did. As the huge dump trucks rolled past with their cargo of death, the ground shook beneath my feet. It was all deeply and strangely hypnotic. The air burned my nose; it was acrid with diesel fumes and so much more. The men at the site were working 12-hour shifts in these conditions. It was horrible and yet I couldn't, and did not want, to leave.

The enemy was no longer here. All that was left was 16 acres of hell, washed in blood and blackness. There was no music to be written about this place that trembled in such terrible agony; and even if it were possible to capture something so horrific, who would want to hear it.

Why had I been called to come here?

TOLEDO, OHIO
1951

By the time I was born in August of 1951, my parents and three siblings were packed into a small apartment in a rooming house with a shared bathroom in downtown Toledo. A strange assortment of misfits resided in the other units, including a couple from the South, a 40-year-old man with a wife he had married when she was only 14.

I was six months old when Mom left me with this couple while she attended a wedding in Chicago with the rest of the family. She didn't want to deal with two babies in diapers. She returned a week later, after which I would go berserk if she tried to leave me anywhere.

I became a continual problem for her; clingy, fearful, painfully shy. But she felt no obligation to meet my needs. She had a deep aversion toward physical affection, specifically toward a female child, and she would never allow me to get close to her.

Unable to understand the abandonment or the rejection, I had to trust there was good reason for it. My mother would not do this to me unjustly. The primitive thought processes of a child could only conclude that there had to be something very ugly, very evil inside of me that deserved this treatment. If I could just fix it, Mom would love me again. The model I would choose to emulate would be my fairy tale characters. There, goodness always overcame evil. I would become so good, so perfect, that Mom would take me back.

In 1954, we moved out to her father's 52-acre farm in Monclova, located in one of the poorest counties of Northern Ohio. Now with six children to care for, my parents settled in as a distant aunt sent over a horse, a pony, and a collie; cats, chickens and ducks were also added to the mix. Mom loved the Banty chickens, so they were allowed to live in the house and roost under the kitchen sink. When the family cat attacked and killed one of them, Mom ordered Dad to shoot the cat. In front of us. Fleas were everywhere. I would wake up in the morning with hundreds of them jumping on the floor, in my bed, and the itch from the bites would drive me half crazy, leaving open sores on my legs bleeding onto my socks.

The only relief from the fleas was the extremely harsh winters ushered in by the winds off Lake Erie. Ice would layer itself on the inside of my bedroom windows, and I would sleep with my clothes under the covers and get dressed before getting out of bed in the morning. One winter was so cold that the pony froze to death.

We lived a strange form of poverty. There was a constant anxiety over the cost of food, but there would always be a three-gallon container of chocolate ice cream for Mom's consumption alone. She was obese. Often, breakfast was a bowl of chocolate ice cream that I quietly stole from the basement freezer. It was rare for Mom or Dad to get up with us; we were left to make our own breakfast and lunches for school. Dad was always worried sick about the bills, but he had the most up-to-date television set and the largest antennae in the county. I assumed financial responsibility early, making the decision to leave the Brownies. No one asked why, but after the first meeting I was certain we could not afford the weekly dues; a nickel.

We were heavily dependent on rain to fill the shallow well, which barely met the basic water needs of our eight-member family. Consequently we were always dirty. I had two sets of clothes which I alternated throughout the school week, and often were the times when I would search through the dirty laundry to find the least objectionable panties to wear another day.

My father's income was not sufficient to support the family. It was suspected that Mom manipulated her father into offering handouts to pay for the whole mess. He died in 1959, and with his death so died her financial security. The horse was given away, the chickens and ducks disappeared. Now, with little distraction, her six children became psychological pawns. Two of the boys were set up as her favorites; the rest of us were duly informed that, to her, we had no value. So became our social structure, our skewed version of the world. I resigned myself to the lowly position of least-favored, knowing my existence made her unhappy, hoping someday I could make it right again. At the same time, she hated my lack of confidence, pushing me into doing things I was terrified to do, becoming furious when I dissolved into tears. When Mom got angry it was like a crazed pig-eyed monster coming and coming and coming at you, and there was no place to escape. It would affect every cell in my body, twisting them into knots of fear; I would do anything to avoid the onslaught.

By the time I was nine, my responsibilities included regularly cooking dinner for the entire family, and washing the dishes afterward. I would shrink before the daunting task of having to cook the perfect meal, since a mistake would unleash a petrifying tirade. God forbid the chicken burned, or I forgot to add the baking soda to the cake and it had to be thrown away. I had wasted food and money which – in our world – was the most terrible thing one could possibly do. I had failed her again.

"Can't you do anything right?!"

If you did something brainless, she would tell everyone about it for days, weeks, and years afterward, while you stood there.

I learned to never question or talk back. I tried to live under the radar, invisibly. I moved without moving. I learned early to never desire or ask for anything, because it became too hard to live in a state of want and need. I denied the very concepts. I became lost in the distortion and finally disappeared altogether, developing an alternate reality for survival's sake. I was a total fake, a caricature of my self, borne of the ultimate humiliation: My mother did not love me.

GROUND ZERO, NEW YORK CITY

I stood staring over the police barricade on Church Street for a very long time. It was not much of a view; one could only see a small sliver of the area from any vantage point other than the viewing platform. And what I wanted to see – the pit where they were still digging – was blocked off by a black tarp on the fence surrounding it. There were still piles of twisted steel beams at the base of one of the buildings in my line of sight, but that was all. I listened as visitors spoke to a policeman to ask where they could get tickets for the viewing platform. His name was Weber. He had a kind face, and impressed me as being very intelligent. This was my first experience of a NYC police officer, and I couldn't get over how nice he was to everyone. His interaction with the tourists always went like this:

Tourist : Can you tell me where the viewing platform is?

Weber: It's just up this block and turn right past the church, but you have to have a ticket to get on the platform.

Tourist: Where do I get the ticket?

Weber: You walk down Fulton street, just past the church, all the way down to the Pier 17. There's a kiosk, you can get a ticket there.

He went through this over and over and over. I was amazed at his patience and his genuine warmth towards everyone he spoke with. I felt a need to speak with him too, to acknowledge the space we were in, and so I simply said it was "unbelievable." He asked me where I was from. Boston. Did I lose anyone here? No, I said, I had come to pay my respects, and to do some research for a memorial piece. I told him I was having a difficult time wrapping my brain around this.

A group of tourists walked up. "Can you tell us where the viewing platform is?"

Weber began again, as kindly as though he were hearing the question for the first time, "It's just up this block and turn right past the church, but you have to have a ticket to get on the platform. To get your ticket, you walk down Fulton street, just past the church, all the way down to the Pier 17. There's a kiosk there. They'll stamp your ticket with a viewing time so you don't have to wait in line all day. You'll probably get a ticket for ... what is it, about 12:30 now? You'll probably get a viewing time for about 3:30."

I interjected. "Excuse me, officer, but I was down there an hour ago, and my ticket is for 6:00 tonight." He made a face, and turned his attention to the tourists. "Then there probably aren't any tickets left for today, you'll have to come back tomorrow."

The tourists thanked him and left.

Weber turned back to me. "Thanks for telling me that, I'm going to have a lot of people mad at me for sending them on that long walk for nothing."

"I was surprised, too. I was expecting to get a ticket for around two or three. My train leaves at 7:30 tonight, I don't see how I'm going to get on the platform and make it back to Penn Station in time."

"Well, there's nothing really to see. They've dug so deep into the ground now that when you're on the platform you can't see anything anyway. Besides, you don't want to go up there. I've seen people standing with their backs to Ground Zero, big smiles, somebody takes their picture ..."

"No! There are signs all over saying no cameras!"

The officer shrugged. "They just do it anyway."

"I just can't believe anyone would do such a thing. It's so sacred here."

"You get all kinds."

"Where were you that morning?"

"I work in the Criminal Investigations office about a mile from here. When the first plane hit, we all wanted to come down here, but they ordered us to stay put, which probably saved my life."

I asked him if he could clarify a few things I had heard, like, why did they "suspect" Mohammed Atta was flying the plane? He told me that "if you watch the video in slow motion, you can see the front of the plane come through the building. The cockpit was found down in the rubble with Mohammed Atta strapped into the pilot seat." He also told me that the fire, which the media had reported as being out, was still burning. That it was not jet fuel, but rather 60,000 gallons of diesel fuel stored in one of the buildings for emergency purposes.

A young man walked up to us. "Excuse me, can you tell me where I can get a ticket for the viewing platform?"

Weber went through the litany again, but this time he added: "But no point going down today, this woman was down there an hour ago, and her ticket is for 6:00; there probably aren't any more left. You'll have to get there early tomorrow."

The man looked crestfallen. "Thank you, officer." He started to walk away, but I reached out and touched his arm. "Wait a second." I turned to Weber, "You truly can't see anything from the viewing platform?"

"Not really."

I reached into my purse and held out my ticket to the man. "Here, take it. I can't use it. My train leaves at 7:30, and I'll never make it through the line in time."

The man hesitated. "Are you sure?"

"Go ahead, take it." There was an acutely painful finality as I handed it over, knowing I would not be able to afford another trip down here again anytime soon.

Weber said something I couldn't understand, and I looked at him quizzically. "What did you say?"

"I said, 'How nice.'"

"Oh. I thought you said 'how much?'"

We all laughed, and the man said, "What, you two got a scam going here, you tell me there are no tickets, and then she produces them and you split it?"

"That's a good idea, I'll have to start doing that," said Weber.

Thanking me once again, the man left.

Officer Weber waited until the man was out of sight and then pointed his finger at me. "You. Come with me." He spoke to the other officer. "Hold the fort, Scott. I'll be back in a few minutes."

Pulling aside the blue sawhorse barrier, Weber led me down Church Street in front of the viewing platform and up to the open gate that led into Ground Zero. It was as though he had ripped away a barrier from around my heart, his kindness almost violent set against the landscape in which we stood. The dismal current of sound faded, and for a split second I thought I saw an invisible glowing path laid out before me, surrounded by the warm peaceful presence that was now becoming eerily familiar.

We did not go further: Hardhats and protective clothing were necessary to go down into the site, and one would have had to go through decontamination to get out. He told me about the layout of the site, where the buildings had been, where we were then; looking at the subway entrance. Where the planes had hit, what direction they had come from, and why the remaining buildings were covered in the black drape. He had interviewed the eyewitnesses in the surrounding buildings who had seen hundreds of people jump. People hugging each other and jumping; ten people holding hands and jumping; people on fire, jumping. As he told me this, I had the oddest sensation that he was talking to somebody else he was seeing in me, but not me ... not me.

There was a remarkable sense of completion to this moment, and I knew it was time to move on. I told Weber I was a massage therapist (it was just too complicated to explain Alexander Technique teacher), and asked if it would be alright to go to a firehouse and knock on the door to offer my services. He said he thought it would be greatly appreciated, and told me of a little firehouse a couple of blocks away. I hugged him good-bye, and thanked him for his kindness.

Although I had not stood in the very middle of Ground Zero as the "voice" demanded, it seemed rather significant that I had made it as far as I did. The warm presence was comforting, and was an indication that I was on the right path. I had received further information. I knew that I was to visit a firehouse, volunteer the bodywork, and return home.

I set out with some confidence, but I missed my turn, and ended up walking down deserted streets near the FDR Drive. I never saw so many excellent locations for a rape and/or murder in one block. One lone man was walking toward me. I panicked; my usual reaction when any strange man approached. I would have to ask for directions, something I was told *never* to do in NYC. (Never let on that you are lost, never look people in the eye, never ask for directions.) Would he rip my purse off my shoulder and run away with my train ticket? Would he pull out a knife and start stabbing me? Was this it? Was this the scene of my death? No, instead he listened and pointed across a large parking lot.

What was wrong with these people? Was I in New York?

Finally, I saw the firehouse. It was so small. The tired and torn flag, the flowers, the few candles burning, the veil of grief hanging over the building; an energy that quietly pleaded, "Please, leave us alone." A searing raw pain emanated from the red door. I read the names of the men who were lost, written on a plaque that hung next to the window: Lt. Robert McMannis. FF. James Gluntz. FF. Elias Keane. FF. Sean Michaels.[1]

---

[1] All identifying information has been changed to protect the privacy of the firefighters and the families who lost loved ones on 9/11.

It was too much. I walked away, crying, hoping no one had seen me. I continued up the block, trying to pull myself together. I walked back, and pretended to read the names again, but this time I noticed a sign on the door: "If you are in trouble, just ring the bell, any fireman will help you." For some reason, those words seemed directed at me. I was in trouble. I needed help. The tears were now uncontrollable, and I walked away again.

This was sheer hell. I couldn't hold myself together. Watching this on television was a whole lot easier than getting involved. Did I have any right to intrude on these men? They wanted to be left alone. How dare I believe I could offer them anything? But if I didn't knock, could I go back to Boston and live with myself, admit that I didn't even try?

I had not brought enough tissues for this. It was now or not at all. I dabbed at my face with the last tattered remnant, hoped my nose wasn't too red, walked back and rang the bell.

With the ring of that bell, everything started to shift, became surreal, like I had stepped into another dimension that was closer to me than my own breathing. The eleventh day of September was no longer safely encased on a television screen in my apartment; it was alive in front of me in the face of the firefighter who opened the door.

MONCLOVA, OHIO
1964

Adolescence was a nightmare. I inherited my father's intense looks, but what I saw in the mirror, I hated. I misread the stares coming from strangers, kids at school, and most everyone, believing they saw only what my mother did; an inner ugliness and evil. Thoughts of suicide kept me in balance, calmly assuring me there was a last-ditch solution. I remained isolated at school, eating lunch alone in the band room. I just couldn't tolerate the stares I elicited by simply walking into the cafeteria.

My escape was to write an imaginary life for myself.

The Beatles landed in America, found me, and thought I was very special. They took me back to England with them, and I lived with John and Cynthia, properly chaperoned at all times. They all loved me and cared very much for me. I was so happy. We wrote a hit song together. We recorded it and I became a singing sensation, appearing on the Ed Sullivan Show in New York City. I married Paul and we had four children, although there was a plane crash and he went into a coma and died and I ended up marrying George. Those were such happy times, and they sustained me.

I left for college, innocent as a lamb and ripe for slaughter. There, I was exposed – without any guidance or protection whatsoever – to hundreds of men. After eighteen years of neglect, the attention was staggering. Believing each was a gallant prince seeing past the flaws and into my determinedly good soul, I readily opened up. Would this one marry me? If so, my true value would finally be revealed; to my mother, and to the world.

There was instead a relentless onslaught of sexual pressure. I had no idea how to deal with it except to acquiesce. Accustomed to giving in to my mother's wants and needs, I had no awareness of what I did or didn't want. Feeling guilty and responsible for "turning them on," I was therefore – somehow – obliged to have sex. And so I did.

Desperate for the physical affection I never had, relationships became a subconscious business arrangement. If I give you my body, will you love me? Will you marry me? Of course I became pregnant. Abortion had just been legalized in New York. I went alone. It was the worst and most painful experience of my life, though there was no moral question. Quite simply, I hated myself. I had no value, and therefore, nothing inside me did either.

The man who eventually married me was 40 years old and determined to father a child. Knowing I could get pregnant made me realize I had finally reached an important goal. I would be married and safe. I didn't know if I wanted a child, but that didn't matter; the happy ending was in sight.

No problem, I assured him. I could get pregnant. But I didn't. Once again, I fell far short of expectations. He suggested the abortion ruined me inside. He insisted that I must have contracted some disease that left me infertile. Pregnancy was now the only achievement that would render me acceptable; I made a deal with him that I would leave in two years if I didn't succeed.

He, however, had already moved on. He had an affair. Unable to trust him, I left. He found someone else to impregnate, and we were divorced shortly thereafter. For two years I lived in a black chasm of depression.

I blamed myself, took time off, worked harder to overcome my failings, entered into a relationship, and failed again. Determined to get it right, I ventured out to seminars on finding my "self." What I found was that I was a total fake. So began my long journey back to becoming authentic. I set higher standards for my own behavior, and believed that a good man would find me when I finally became truly good.

It seemed, however, that I was unable to rise above the prevailing standard. *The men I encountered felt they had every right to grab me, kiss me, put their hands through my hair, or bring me in for a hug that was ultimately a ploy to feel my body pushed up against them. A client ripped off his clothes and demanded I watch him masturbate. I couldn't walk by the Charles River without some stranger in the bushes whipping out his you-know-what and masturbating while looking at me. It even happened on the way to church on Sunday morning. I stopped going to the river, got a ride to church. In benign social situations, men I had never met before would make specific and inappropriate sexual comments within ten minutes of starting a conversation. My married landlord would propose a menage-a-trois whenever he saw me, and when I moved to another apartment, the next landlord showed up drunk at my door with that "look" in his eyes. I knew what he was after.*

I was hunted, and it was frightening. Yet men refused to acknowledge my fear, focusing only on their sexual fantasies. With my limited self-awareness, I wondered: Was I doing something to invite this? I stopped smiling, wore baggier clothing, stopped wearing makeup, and started gaining weight. It didn't matter. Something was drawing the crudest sexual energy my way. I had no control over it, and it scared me.

The few high-caliber men I did meet were not interested. They didn't ask me out, which further confirmed my fears that my inner evil, so reviled by my mother, was real. Their rejection left me angry, powerless, and ultimately hopeless. I concluded that on some deep, subconscious level I wanted this perverted attention, as it was all I was getting. The realization spoke of something so dark and sickening, I could not even begin to approach it.

Since I couldn't fix it, and no one else could either, I surrendered. Retreating to the safety of my apartment, I rarely ventured out past the necessities of work, groceries, dinner with friends, and movies alone. I was always ready for the next sexual attack, but as long as I stayed away from men, no attack came. As the years went by, my life smoothed out and quieted, proof enough I was better off without them. Good riddance to bad rubbish. September 11th proved to me once again that they were still out there with their insatiable egos and endless desire to destroy anything good; monstrous men doing monstrous things to innocent people.

I must take time now to describe my understanding of firefighters prior to 9/11. Although I had never met one, somewhere I had gotten the idea (probably from my mother) that they were uneducated, beer-drinking, blue-collar workers in a dead-end job. And therefore, I assumed I was going to be intellectually superior to them all.

And they were *men.*

ENGINE 32
NEW YORK CITY

Schultz, it read, on the navy blue FDNY shirt of the firefighter who opened the door. Not exactly the Irish name I expected to see. He dwelled in his mid-to-late thirties, with buzz-cut red hair, a pale complexion, a mustache. He wore navy blue pants, and his shirt was carefully tucked in above a black belt. He appeared to be solid muscle, with an impossibly broad chest and huge arms. I was not prepared for his professional demeanor. My assumption was that these men would be emotionally destroyed by the events of 9/11 and open to anyone coming to help, but I saw nothing like that in his face. A fire engine loomed large behind him and engulfed the entire area, leaving only a small space for us to converse. I had the distinct impression that I was not going to be allowed in, either by him or that big red engine.

"Can I help you with something?" He had a very simple, almost childlike way of speaking.

"I just want to thank you for what you guys did ..." My voice choked as unwanted tears began. He said something comforting in return, trying to reassure me, and I said, still crying, "But I didn't come here to do this." I made a self-deprecating motion with my hands toward my face to indicate that I did not approve of my tears or my weakness before him. I continued. "I came here to do something for all of you. I do this bodywork that's very different and I have five hours until my train leaves, and I'd be happy to work with any of you if you would like."

"We're fine, but thank you anyway."

That was all there was to it.

I was dismissed. I was confused. Who in their right mind would turn down free bodywork? Certainly no man I ever knew. Suddenly, I felt out of my league and oddly uncomfortable. He had relieved me of all further responsibility in this matter, and I could go home, back to the safety of my apartment. I had risked enough for one day, for one lifetime.

Instead, I found myself insisting that this work I did was very different and he should try it. I thought he would just say no again, but he didn't. He simply said, "Let me go check with the boss."

He left me standing next to the huge fire engine; an incredibly complex and beautiful piece of machinery, I thought. It seemed almost animate, making sure I stayed where I was. A couple of minutes went by, and then he returned with an older fireman dressed in the same intense blue, who looked completely exhausted. He glanced at me quickly and said in a flat voice, "Sure, go ahead."

I followed the firefighter up a long flight of 22 heavily-worn black steel stairs. There was a beautiful and intricately-wrought railing that had been painted black long ago, as evidenced by its chipped paint. All along the wall were letters and cards, drawings and plaques from people all over the country. He took the steps two at a time. I struggled to keep up, my 50-year-old, 20-pounds-too-heavy body was out of breath by the time it reached the landing.

At the top of the stairs we turned left through a doorway, and straight ahead was the brass sliding pole disappearing down a wide hole in the floor. We went through a grey metal door into the room where they slept. The uneven walls were painted white, and the ceilings were beautiful molded tin that had been painted white as well. Harsh panels of bare fluorescent lights were suspended throughout. Heating pipes worked their way through various parts of the room in odd configurations; a picture of a vase of flowers was inexplicably hanging upside-down, and half-obstructed a window. All the beds were neatly made, with threadbare sheets and brown wool blankets.

A makeshift office held boxes of t-shirts and sweatshirts stacked to the ceiling that were being sold to raise money for the families of the men who were lost on 9/11.

"I'm Jessica, by the way," I said, holding out my hand. He shook it. "Mark." I pulled up a chair from the desk and had him sit. I asked if he had anything bothering him, and Mark divulged his symptoms. I began to work.

Although it is not part of the Alexander Technique teaching process, I have always been able to visualize -- on a screen in my mind's eye -- the energetic vibrations from, and internal structure of, a student's body, not unlike an X-ray. This ability allows me to be extremely accurate in assessing a student's needs. As I started working with Mark on this intuitive level, something occurred which I had never experienced in my 16 years of bodywork. I felt him intuiting my own needs in the same manner I was intuiting him, but with a great deal more intensity. It was astounding, yet as plain as day: He didn't want the bodywork. Instead, he was letting me work with him to make *me* feel better.

What the hell was this? After all this man had been through, he was now sitting in this chair for me? How could he muster the time or energy to care about my feelings at a time like this? Since 9/11, there had been no place for me to let down my guard and no place in the world where I felt safe. For this brief moment a hand was being offered to me with the assurance: "You are safe here."

It made me feel vulnerable and exposed. I assured myself that this couldn't really be his intent; that I was reading it – and him – all wrong. Men were not capable of this depth of understanding.

I managed to shrug off the experience and regain my concentration. Nobody had ever taken care of me, and I was not about to let this guy start. I had trusted too many men and had been royally screwed over for that trust. But I was more than just a bit shaken by what had just happened.

Although the Alexander Technique is a subtle form of bodywork, Mark was completely aware of everything I was doing. He declared that this was unlike any massage he had had before; that his muscles seemed to be moving and releasing on their own. Fifteen minutes later he stood up, and reported that his pain was gone. His demeanor was markedly different; gone was his ultra-professionalism and there was a relaxed gentleness in its place.

I felt an extraordinary confidence. This is what I had come to do, and I was in control. I said to Mark, "Okay, now go get me another fireman."

He went downstairs, and few moments later up came the "boss." An attractive man in his forties, with grey hair and brown eyes, Harry was in a bad mood, and also seemed to be in bad shape. He sat in the chair and told me of his six months of neck pain, and about the last massage therapist who had been there and worked on him and made it all the worse. "If you don't mind, please stay away from it."

He wanted the bodywork, but he sure as hell wasn't going to trust me. He spoke incessantly – about his wife, his kids, his job, and the men who were lost on September 11th. I took it all in, listening between the lines, and put my hands on his neck. Almost immediately an intense heat rose into my hands.

"Do you have any idea how hot your hands are?" he asked. "Have you ever put a thermometer on them? Do you realize it's physically impossible for your hands to get that hot?" And then, "Are you one of these healer types I've heard about?"

I said, "Yes, but please don't say anything to the other guys, they'll think you're crazy and ... they'll think I'm crazy, too." But it was a relief to be honest. As the session progressed, the trust grew and he quieted down.

In this arena, I knew what I was doing, and I was sure of my abilities. I spent about two hours with him; restoring him back to a self he had long since thought was gone. He stood up straight, his neck free of the nagging pain, and he smiled. His face glowed with appreciation. I was so happy to have given something to someone who had been through so much. I had a sense of accomplishment and purpose that I had never felt before. By taking care of these men, I was fighting back.

I made a decision. A memorial piece of music would do nothing for these firefighters; they needed bodywork, and lots of it. If I was going to do this right – and have a concrete influence on their welfare – I would have to work with them for an extended period of time, at least a year.

I told Harry that although he was feeling better, he was going to need more sessions. I made a commitment to him then and there to travel to New York City as often as possible. I was going to "adopt" Engine 32.

"You don't have to do that, but if you are ever down here, please stop in, we'd love to see you," he humbly responded.

"No, I mean it. I'm adopting this firehouse and I am going to keep coming down here until I know you are all okay." I was scaring myself to say it out loud. This man was not only listening to what I was saying, he believed it.

While we were talking, Mark strode in. Harry raved to Mark about how well he was feeling.

"I know, I can still feel where her hands were," Mark replied, equally positive.

Harry said, "We have to give her something. How about a t-shirt? Mark, do we have any t-shirts up here?"

Mark opened a large green tin cabinet neatly stacked with folded shirts. He handed me an official navy blue Engine 32 shirt with a magnificent FDNY logo silk-screened on the back. I was thrilled, meanwhile protesting that they didn't have to give me anything.

Harry had no reservations about showing his gratitude. "C'mon downstairs, I want you to meet the other guys. Can I get you anything? Are you hungry? Thirsty? What can I get you? We've got ice cream, we've got tea. What can I get you? Anything?"

It was as though he couldn't give me enough for what I had done. Not the usual guy thing.

I followed downstairs, squeezing past the fire engine; still imposing.

The floor was oil-soaked and uneven, with deep grooves worn into the stone from the weight of the engine. There was a basketball hoop over the door to the dining area, and an old black barbecue grill stood amidst the bunker gear hanging on the wall to the right. It reminded me of a college frat house.

Half of the dining room was taken up by two garish recliners and a broken-down couch adorned with two worn quilts. Stained office chairs in various states of disrepair sat around a large rectangular wooden table. A blue and red pen, rubber-banded together with a large, important-looking journal, lay out on the table. On the scarred walls hung awards, bulletin boards, and two large chalk boards covered with scrawled notices of funerals, memorials, fundraisers, and outings.

Harry introduced me to the other guys who were sprawled about like cats, lounging in the chairs and resting on the couch. I had fully expected to instantly register on the "male scale" as I always have, but nothing was happening. I had this odd sense that I was looking up at all of them, as though suddenly I was three years old and in the presence of adults. Something was very different here; what I expected continually failed to materialize. The belief that I was superior to them was fast eroding and, still unsure of the dynamic, I decided the best action was to keep quiet.

Harry pulled out a chair for me and I found myself sitting next to Richie, a handsome firefighter with blonde hair and blue eyes. But it was his energy that caught my attention. He was young, about 32, and had a playful way about him. I would find out later he was an Irish Mormon, an odd combination.

Harry brought me a glass of water, and tried to nudge the other men into a bodywork session. They all declined. Both Harry and Mark again thanked me for my help, and left the firehouse.

So there I was, sitting in a room full of men I didn't know, drinking my water. It was quiet; an unspoken game of "chicken" – who would speak first. Finally, Richie turned to me and said, "Where are you from?"

"Boston."

"How come you don't have a Boston accent?"

"I'm originally from Ohio."

"What made you leave Ohio?"

"I was chasing some guy." Should I have said that? It was the truth. (They were all listening.)

"Did you catch him?"

The way he asked me was so sincere, it took me off guard.

"Well, his parents didn't think I was good enough for him, and they threatened to withhold his tuition if he continued to see me."

"I'm in a situation similar to that myself right now," Richie replied.

I was surprised. My expectation was for him to make a joke about it, and instead there was something about his response that let me know he understood how painful that had been. Again, as with Mark and Harry, there was something more to him – a truly empathic nature that allowed me to let down my guard. How could there be *three* men like this?

The conversation presented an opening for me to talk about the Alexander Technique and so, with a little convincing, he agreed to come upstairs for a lesson. He didn't talk much while I worked with him, but I remember clearly thinking that if I had a son, I would have wanted him to be like Richie. He had everything going for him; looks, intelligence, a sense of humor, and an elegance in the way he carried himself.

Afterward, Richie and I went back downstairs. None of the other men expressed any interest whatsoever. They talked right past me, but strangely, it didn't strike me as disrespectful; they simply considered me unable to understand what they were saying, as though I was a small child among adults.

The lack of sexual energy directed toward me was wholly bizarre; I didn't know how to function without it swirling in the air. My guard was up, but there was nothing to resist. Had I aged so much lately that I was now undesirable? I had a sudden urge to run to a mirror for verification.

Well, I thought, resigning myself to this new reality, meeting a man was not on my agenda anyway. I had a big job to do, taking on this firehouse, and the lack of sexual attention would make my work a lot less complicated.

It was five o'clock, and I needed to catch my train at Penn Station. One of the firefighters was going off duty and offered to walk me up the street and aim me in the right direction. As I edged past the huge red engine on my way out, I resisted the urge to give it a comforting farewell pat. I thought the firefighter would think it stupid.

The cabdriver took a right down Barclay Street, past Ground Zero. I asked him to slow down as we neared the site; I needed to offer a last gesture of respect. I fully expected to again be overwhelmed by the painful vibrations of all that had happened there. But the horror of the place instead bounced away as though my very being was surrounded by a protective shield. I was infused with inner warmth and light, and for the first time in my life I felt no fear ... only the certainty of the goodness of being alive, a completely unfamiliar emotion. I turned around in the seat, staring back at the gaping cavity and the glaring overhead lights until they disappeared from view. I felt no fear.

I felt no fear.

Within the scarred walls of Engine 32 – among mismatched furniture and beyond the big fire engine and the unusual men – something remarkable had happened to me; within me. I combed through every detail of the experience in my mind, searching for the answer. I couldn't find it, but this much I knew: Whatever it was, I had been looking and waiting for it all my life.

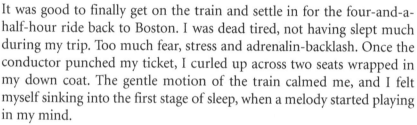

It was good to finally get on the train and settle in for the four-and-a-half-hour ride back to Boston. I was dead tired, not having slept much during my trip. Too much fear, stress and adrenalin-backlash. Once the conductor punched my ticket, I curled up across two seats wrapped in my down coat. The gentle motion of the train calmed me, and I felt myself sinking into the first stage of sleep, when a melody started playing in my mind.

E-F-E-D-C-D-C, and it repeated again, and again, insistent and louder. Finally relenting, I sat up and pulled a pen from my purse.

Not since the immediate aftermath of 9/11 had I heard a melody in my head. Instinctively, I knew what it was. A doctor had shot video of the World Trade Center after the first tower fell and in the background was an eerie and constant chirping sound, echoing in a redundant pattern. Later, I watched that doctor on a talk show as he explained the peculiar audio. It was the sound of the PASS alarm, or "mask," as the firefighters called it. When firefighters are running low on air, or if they haven't moved for thirty seconds, this alarm will activate. In the video, hundreds of these alarms were going off simultaneously. The pattern of tones emitted was incredibly haunting, and I made note that if I ever were to actually write the memorial piece, I would interject that pattern into the composition.

But here it was, repeating in my mind, creating an entire piece unto itself. I wrote the notation on a scrap of paper. My mind quieted, and I finally fell asleep with the Engine 32 shirt clutched in my hands.

JANUARY 14, 2002
BOSTON, MASSACHUSETTS
12:30 A.M.

The notes evolved into the story. First, a simple melodic statement of grief and mourning; string trio. Then, the introduction of *Amazing Grace* with English horn, French horn, and trumpet. In they came, separate and alone, a nod to the fragments of countless funerals I witnessed on television. Rising, a crescendo recognizing the dedication of these men to their work as the simple melody was now performed by the entire orchestra. Next, the introduction of *Where, Oh Where Has My Fine Young Laddie Gone?* – with high strings and oboe solo symbolizing the crying of the widows and children left behind. Finally, all of the forces of the orchestra gather in an audacious march to denote the bravery of these men as they faced their destiny on that fateful September day.

My hands played the notes without the assistance of my brain, like someone was controlling them. They deftly expressed the emotions of the piece beyond the realm of what I could have consciously created. In half an hour, the piece was sketched, not so much by me as *through* me.

A few days later, I wrote a thank-you note to Harry, Mark, Richie and the others, thanking them for my Engine 32 t-shirt. It hung on the back of a chair in my studio where I could see it in every waking moment.

January was generally a bad month for work – client or music – and my finances were looking grim. Another trip to NYC was out of the question. There was nothing for me to do but finish the orchestration of the piece (assigning which instruments would play each note), which I completed over the next few weeks. Now, I was challenged with adding the names of the 343 deceased firefighters in voice. But I did not intend to sing each name as though reading from a list, but rather, to unite them in a blended harmony as a symbol of their brotherhood. Individual names would not necessarily be heard. Their names – representing their individual energy, their last earthbound light and the reflection of their souls – would hopefully project my psychic experience of them at the firehouse. Would it work, or sound like mumbo jumbo? I would have to sing and record them all to find out.

Peter Ganci, Jr., Raymond Downey, Mychael Judge – I was not prepared for my heart-wrenching reaction. I sang name after name, and found myself breaking down over and over. The list was seemingly endless. The worst moment came when I read the name of Joseph Angelini, because the next name was Joseph Angelini, Jr. I was sure it was a mistake. His name must have been noted twice. I researched it. Father and son.

*　　　*　　　*

On January 29th I received a letter from ASCAP, an association that collects royalties for musicians whose music is aired on radio or television. I almost collapsed on my front porch when I opened it to find a check for $3,000. Music I had written ten years earlier had been used on cable television, and I was now being compensated for it. I would be able to keep my promise to Harry and return to Engine 32. The warmth of that promise hovered around me, bringing a huge smile to my face. This was no coincidence. I was meant to go down there as ordained by this angel, or whatever it was watching out for those men.

The money should have gone toward a new car. My poor Audi Quattro was 17 years old and I had waited years for such a lump sum as a down payment. But what was going on in NYC was more urgent, and keeping my promise to Harry was of supreme importance. This commitment to Engine 32 would no doubt cause some measure of financial hardship, but for the first time I realized that I knew how to be poor. There was an odd sense of power associated with that knowledge. There are many levels of poverty, and over the years I had mastered most of them. I knew how to squeeze and put off and tell landlords that they would need to wait for the rent. Always, I had landed on my feet. It was ironic that all I had seen as failure in my life was now transforming into the very strengths I would need to work with Engine 32. The ability to stay separated from my feelings would be essential to working with these men without falling apart. In comparison, to make do with little or no money was a piece of cake.

I called Engine 32 for the first time. I was very shy and very excited as I asked to speak with Harry. He was genuinely happy to hear from me, and even happier to hear that I was planning a trip down for four days at the end of February. I told him the miraculous story of the ASCAP check. He was a little worried that I was availing myself of both time and money with no guarantees that the other men would accept the

Alexander work from me. He offered, "I'll try to talk to the other guys, but I don't know if they'll go for it."

I immediately reassured him: "Harry, if I end up working only with you, it will still be a worthwhile trip for me. However it goes is how it goes."

MONDAY, FEBRUARY 25, 2002

As the Amtrak train drew closer to New York City, I began to recognize that the sick feeling in my stomach – the fear – was a necessary accompaniment to the act of living; a consequence of moving forward without knowing the outcome. This plunge into the unknown would not necessarily yield a bad result. Conversely, trying to hold my life still – not moving or growing – had given me an illusion of safety, which in fact, had never existed at all.

MANHATTAN

Hotel check-in wasn't until 2:00, so I stopped at a payphone in Penn Station to call Harry and tell him I had arrived. He said, "Come for lunch. Come here right now!"

I wasn't sure if I could handle sitting at a collective table with all my dietary restrictions in tow – no red meat, no dairy, no wheat, no sugar, no caffeine. I hesitated. He sensed it.

"What are you, a vegetarian? Are you one of those vegans?"

"No, but ..."

He interrupted me, "We're having tuna melts; can you eat tuna? Can you eat melt?"

He was so direct and open, I had to laugh. He continued. "It would be good for the guys to have a chance to get to know you, break the ice, you know what I mean? C'mon, have lunch with us." He was right. My spirits began to lift out of the fear.

The cab drew up in front of the firehouse. I rang the bell, and a tall, gorgeous and muscular firefighter answered the door. There was no guarding of the door today.

"Hi, they're expecting you in the back. Let me take your luggage. I'll carry it upstairs for you."

"Oh, I'm not staying here, I just can't check into my hotel until two."

"You're welcome to stay here if you want."

Whoa, I thought in amusement. When I'm in a nursing home all alone with my memories will I regret that I didn't take him up on his offer? Harry came to the front on the narrow path past the fire engine, a big smile on his face, clunking along in his unlaced shoes. After a warm handshake, he said (all in one sentence as usual): "C'mon back, the guys want to meet you, did you meet Scott? You should work on him, he's in a lot of pain. You should work on him first. C'mon, we saved lunch for you."

Everyone else had finished eating, and were up and about, doing dishes, cleaning off the table. I was briefly introduced among the hubbub. There was Scott, the handsome one; and Manny, a very quiet and shy personality with kind eyes. Lou with his shaved head, pointed features, and military demeanor, giving me a curt nod; and Ray, a wiry, tough, dark Italian with a mustache who was all cold business and looked like he wanted to stay as far away from me as possible.

Harry sat and watched patiently while I ate tuna and cheese on an English muffin and a salad. He pointed to the chalkboard where he had written my name and *Alexander Technique*, along with the dates I would be available. He again insisted that I work with Scott, but Scott said he would be working that evening, too, so he insisted that I work with Harry first. I was touched that each would defer to the other.

Upstairs we went. I asked if we could stack a few mattresses on a bed to create a worktable of proper height. I was stunned at the poor condition of the beds. The mattresses were lumpy flat cotton set on a metal frame with springs which collapsed into a "V" when sat upon.

"Prison mattresses?" I said to Harry. "You guys actually sleep on these things?"

He replied that they didn't get much sleep anyway on account of the alarms, and that most of the guys slept downstairs on the couch or in the chairs.

The smell of fresh paint permeated the room; the men were renovating to include an office. Scott stuck his head in the door and asked if I minded the loud rock music. I said I did. He recoiled at the bluntness of my response, and I immediately felt remorse that I had not responded more tactfully. This was, after all, their firehouse.

I started to work with Harry, and he began to question me. Was I married? Did I have a boyfriend? Why wasn't I married? I smiled in an attempt to sidestep the inquisition.

"Do you think if I had a boyfriend he'd let me come down here and be with all of you?"

I was not accustomed to relinquishing control of my working environment, and being asked personal questions was not something I would normally encounter nor tolerate. But here I knew that to hold myself up in a superior position would create awkwardness. To not allow him to know me might risk losing him, and ultimately, all of them. The little bit about these men I had observed thus far was that they were extremely direct and honest. Most had never had bodywork, and were uncomfortable receiving hands-on care, especially in the macho environment of the firehouse. Asking questions allowed them a modicum of control, and so I gave that to them.

Because he asked, I gave Harry a brief story of my marriage. And then we were suddenly interrupted.

BING BONG!!!!!! "ENGINE!!!!!!"

I jumped two feet from the shock of the extreme volume. Harry laughed and leaped up, saying they had a "run" (alarm). I panicked.

"Should I leave? What should I do?" How would they trust a veritable stranger in their firehouse?

Harry responded as though it were the only logical answer, "You're coming with us."

"WHAT?!"

"C'mon, I'll meet you downstairs." He dashed to the pole and slipped out of sight.

Should I go with them? What was the risk here? I had not heard of any woman being raped or killed while riding in a fire truck. Did I want to go? Yes. Maybe. Yes. Please don't let me do anything stupid. I ran down the stairs.

The fire engine had wakened with a ferocious, visceral roar that shook and echoed throughout the building. Red lights flashing; big red door rising. The driver and Lou were sitting up front. Harry was pulling on his bunker gear, and he handed me a firefighter coat to wear. It was so heavy I nearly dropped it. He helped me into the coat, and motioned for me to sit directly behind the driver facing the back. The step was exceedingly high, and I failed to negotiate it on my first attempt. I was so out of shape, and the coat was heavy. How clumsy did I look, and who noticed?

The engine pulled out onto the street and paused. Ray was stopping traffic, and Harry was hitting the button to close the red sliding door. They both got in with a resounding slam of the doors and we were off; sirens blaring and horn honking to push the traffic aside as we made our way down the narrow street. With the excitement of a child running through me, I clung to the steel handrail mounted on the engine housing as I was tossed and jounced about.

Engine 32 carried 500 gallons of water, and you could feel the extreme weight of it as we moved through the streets and around traffic in fits and starts. Up to Broadway, and down Vesey Street past Ground Zero. We were on a glorious mission. The roar of the engine made it difficult to talk or to hear the talk of others. Harry was laughing at me; the look on my face must have been priceless. "OH MY GOD!" I mouthed to Harry. No wonder these men did what they did. This was … fun!

Ray, cold and reserved, sat diagonally across the big center hump that housed the huge diesel engine. I looked over and caught him looking at me, expressionless. I felt bad to be enjoying myself this way among men who had faced so much, but I could not hide the fact that this was truly the most extraordinary thing that had ever happened to me. I was in a fire engine, with five firefighters of the FDNY, in the streets of New York City. It was insane and it was sheer bliss. My former life had been obliterated and I was starting over again, right here, right now, in this fire engine. These men didn't know I was broken, and they didn't have to find out.

Harry took his helmet and put it on me. It weighed a ton. A female cop saw me from the street and gave me thumbs up. I waved back.

As Harry shouted out landmarks above the din, I twisted and turned in my seat to look out from every vantage point. Everything was so exciting from inside this cab; I was seeing the world new and afresh. According to Harry, Engine 32 was one of the first to respond that morning to the WTC. The engine was crushed under the North Bridge walkway; what we were riding in now was an old spare.

I reached into the pockets of the heavy coat, curious to know what a firefighter carries around with him. Kleenex? Chapstick? No. Inside the left hand pocket was a gigantic wire cutter that looked like it would cut through chains, and it must have weighed 20 pounds. I pulled it out and showed it to Harry. We laughed and laughed at the unintentional joke on me.

The fire engine came to a stop. Harry jumped out, grabbed a length of hose and asked me if I wanted to get out. No. I was only too happy to be right where I was at that moment.

I sat there alone trying to grasp what was happening. A new version of myself was being born in, of all places, a fire engine. I could not deny it, feeling myself being lifted up higher than I thought it possible to exist. It was magical. It made me laugh and cry at the same time; laughing to be so easily lifted out of the pain and ugliness that my life had been, and crying for the 343 firefighters who had lived their lives filled with this joy; riding just like this on that beautiful morning in September.

They had lived more life in ten minutes than I had in fifty years. Suddenly I did not feel worthy to be with them. There was something about this job – the willingness to risk one's life in sacrifice to others – that attracted truly great men. Or, I thought, was it possible that the job took ordinary men and *made* them great? No matter which, I was reduced to a state of humble awe. I didn't deserve to be here.

The firefighters returned, storing the hoses, air tanks, and tools – clunk, clank, clunk – and climbing back into the fire engine. False alarm.

Matter of fact, businesslike, no big deal. I stared at them, these seemingly ordinary men, but now I knew they were extraordinary.

We returned to Engine 32, slowly now, flowing with the traffic, but I did not want the ride to end. We pulled up in front of the firehouse to execute backing into the building. Harry told me to "stay put." With the exception of the driver, all of the men got out to halt traffic and open the door. It was a beautiful dance. They held the power to stop the whole world as they returned the engine to its place; each man taking their position, knowing their job. The loud beep, beep, beep, as we backed in, echoed as the acoustics of the huge garage took over.

Harry opened the door and helped me down. "Well, what'd you think?" The smile on his face let me know the question was rhetorical; he already knew the answer.

I beamed. This was no dead-end job. This was the best job in the world.

We went back upstairs to resume his session. Within ten minutes: BING BONG!!!!!! "ENGINE!!!"

Harry leaped up. "Are you coming?"

"Can I?!"

Oh boy! I ran down the stairs, this time feeling more confident that I knew what to do. He didn't make me wear the coat. I was grateful. I was able to climb up into the rig with a bit more polish.

Again, it was a false alarm, but who cared? We came back, went upstairs and started up again.

BING BONG!!!!!! "ENGINE!!!"

This time I rolled my eyes. I wondered if I would be able to get anything done. Harry asked me if I wanted to go, and this time I declined. I needed to save my energy. I had envisioned working with at least four men between 2:00 and 8:00, and the plan was disintegrating with each subsequent alarm.

They left with the roar, the ringing of the doorbell, and the big door sliding down. Then, silence. I was alone in the firehouse.

I was amazed at the level of trust they already had for me. I lay down on the bed to rest, knowing I was operating on adrenalin fumes and very little sleep. I stared up at the ceiling and was comforted by the pure whiteness of the old tin. The environment possessed an odd energy, and again I felt surrounded by love, as though I was being washed clean of my inner ugliness and evil. The walls seemed almost radiant with the energy of those who had been here. Whatever it was that was so wrong inside me, it would not be permitted to exist within this firehouse. If I could only stay here long enough to have it exorcised forever.

It didn't take long for the guys to return; again, the big door sliding open, the familiar beep, beep of the engine backing into quarters. I got up and ran to the stairs to watch from above. No matter how tired, I will not show it. If they could work 12-hour shifts at the World Trade Center site under such grim conditions, I could manage to get through the rest of this day without crashing.

Harry came back up, and we were able to finish. I told him he needed one more lesson with me before I left on Wednesday. We said our good-byes, and he called for Scott to come up.

Scott sat down on the bed and winced in pain. His rib was bruised or broken but he didn't want anyone to know. If he went on medical leave, somebody else would have to cover for him. With 343 men dead, hundreds more out on medical leave, everyone was exhausted and overworked, and so they all tried to stay on the job, injured or not, to carry the load evenly.

Scott was one of the most physically beautiful men I had ever seen; God must have personally done the drawing. Everything was in perfect balance. Not that it mattered to me; in fact, it worked against him. I'd known too many of the pretty boys of Harvard who used their looks to score with women. I was grateful to now be old and wise; 30 years ago this guy would have ended up as just another bad memory.

However, my prejudice against him fell away as he stayed completely genuine during our session. And as we moved from chit-chat to the physical problems at hand, he became quiet in his trust for me. It was what I had come down here to do; provide a safe place for them to let down. His trust made me feel special and whole.

A call came over the intercom. They were going out for the meal, and Scott sat up. "Are you staying for dinner?" he asked. I was relieved since I had no clue of where to go for a meal in this neighborhood. I realized that I was very hungry. He asked me if I wanted to go with them to get it, but I declined, knowing I should still try to grab a nap if I could. He left.

Again, I was alone in the firehouse. I took the opportunity to use the bathroom which adjoined the bunkroom. It was immaculate. I wondered if I should put the seat back up. Out of respect, I did. Looking in the mirror I was stunned to see my face. I looked completely different. I looked genuinely happy. *Who is that?* I wondered.

It's a funny thing, to be in a place where there is nothing that connects you to your customary life. Everything here was different … like being on another planet. Here was this funny little house with five men working together in many ways; cooking, making the beds, cleaning, firefighting – like lost boys of Neverland. How could men live together like this in such harmony? I couldn't live with even one and achieve such a peace. Why?

I heard the engine return. It was shift time, and there were a lot more firefighters there than had been earlier. I came downstairs and walked past the fire engine as the men were getting out of their gear. I felt so shy, and as I entered the dining room/lounge area, I was immediately drawn to the elegant presence of a lone firefighter of African descent sitting at the table. Years of hard work lined his face. He held unquestionable authority, and dignity. I began assessing him on the psychic level, and found him doing the same to me. Neither of us spoke, and before we could, the room was awash in blue as it filled with more and more firefighters. There was a warm chaos in the kitchen, with smoke from broiling steaks permeating the air, and the sounds of dishes and pans clanging and banging, men talking, the phone ringing, and *Jeopardy* blaring on the television.

The table was magically cleared, and a pile of silverware, knives, paper towels, salt, pepper, ketchup, and steak sauce replaced the day's debris. Six plates were laid out. Someone announced "chow" over the crackly intercom.

The lone firefighter was sitting at the head of the table. I sat to his right, and Scott sat next to me. Scott briefly introduced me to everyone. The men were shy, but so was I.

A platter with a pile of well-cooked steaks was brought to the table. "You eat meat, don't you?" asked Scott.

"Yes, occasionally," I lied, and he laid half a cow onto my plate. Garlic bread, roasted potatoes, and cooked onions found their way to the table. Everyone dug in, piling their plates high. We ate in relative silence.

The lone firefighter said something, and Scott offered, "Now, Elliot always mumbles when he talks, I can never hear him, and when I do hear him I can't understand a word he says, so you're going to have to translate for me, okay?"

He was treating me like a little kid, giving me a sense of belonging by giving me a job to do. I loved it. The drone of the television filled the empty air, and we all looked up occasionally to watch.

Scott asked someone to pass the Pepsi, and as he did, I happened to catch Elliot's eye. Elliot looked at the Pepsi, looked at me, and nodded.

Following instructions, I turned to Scott and said shyly, "Elliot said he would like some Pepsi."

Scott's mouth dropped open in amazement. "When did he say that? WHEN DID HE SAY THAT?! I must be losing my hearing; did any of you hear that?" Everybody laughed.

And then Elliot spoke, loud enough for everyone to hear: "So, what group are you associated with?"

Everyone was listening.

My heart sunk. It had never occurred to me that I needed to be authorized to be here. I felt sick inside, and my face turned red. I said, "I'm not with any group."

Elliot's expression remained the same. "You're here by yourself?"

"Yes," I said, full of guilt.

"You from around here?"

"No, I came down from Boston."

His face scrunched. "You staying with friends?"

"No, I'm staying at a hotel."

He pondered this for a few seconds.

"By yourself?"

"Yes."

Silence.

"Are you seeing any sights, any shows, while you're in town?"

"No, I only came down to work with you guys."

Silence.

Elliot said, "You spent your own money to come down here?"

"Yes."

There was another long silence while he took in this information. He finally said, "So you're down here just for us?"

"Yes."

Elliot shook his head in disbelief, while a wave of emotion vibrated through the room. Across the table I saw tears in the eyes of the firefighter sitting across from me. I felt something in the air directed at me, the likes of which I had never felt before; I couldn't even describe the sensation.

The sound of the doorbell diffused the moment. Manny got up to answer the door. Scott explained to me that there were guests arriving. Privately I wondered if I was going to be able to give any more sessions.

A heavyset woman and her teenage daughter came into the dining area. The men were very gracious, offering them their seats, and dinner, but the women declined.

BING BONG! In 30 seconds the men were gone.

The woman made the introductions: Linda Johnson, President of the Santa Fe Visitors Bureau, and her daughter Abby. Linda told me her group was sponsoring an all-expenses paid week-long vacation to New Mexico for Elliot and his wife. They were here to meet Elliot personally. I would find out later that a lot of organizations were making similar offers in order to give the firefighters some relief and distance from the trauma of 9/11. Linda mentioned that Elliot was taking them on a tour of Ground Zero that evening. In that instant, I knew that this would be my opportunity to stand at the center of the site – my instinct to do so was still strong.

The men returned, but they did not back the rig into the garage. Elliot came in and said to the women, "Your limo is ready, ladies." I looked directly at Elliot and blurted, "Can I go with you?" I could tell he was surprised by the sense of urgency in my voice, but he must have also respected it.

"It will be a tight squeeze, but I think we can manage."

The two women were helped through the side door into the back. Another firefighter by the name of Lou hoisted me up into the front seat, which was as high up as the top of my head. Part of what I loved about the fire engine was that the hugeness of it made me feel like a kid. So, there I was, squeezed up front between Lou and Elliot and the onboard computer. The windshield was immense, and looking out once again at the streets of New York City perched so high above them was exhilarating.

As Elliot drove, he shook his head and said, "I just can't believe you came down here all by yourself."

"Elliot, it was the least I could do. I couldn't stay up in Boston and not do something." He repeated his disbelief a few more times, forcing me to wonder if he actually knew on a psychic level how difficult it had been for me. Clearly, for Elliot, just my showing up was enough to earn his respect.

Linda's teen daughter was wearing perfume, which now permeated the interior of the truck. Elliot hollered to the back, "Mmm, mmm, say boys, do you ever remember this fire engine smelling so good?"

I was surprised to find myself feeling a little jealous. I made a mental note that I would find a nice perfume to wear for these men; certainly it was justifiable as a form of aromatherapy.

We drove into the site from the far side, an entrance on West Street, and zigzagged our way in amidst a series of huge dump trucks entering and exiting on the same thoroughfare. Before us lay a huge open black pit, a basin dug deep into the ground where the excavation work still continued. Elliot brought the engine to a stop. Lou opened the door and got out. Offering me his hand, he warned me that it would be muddy. I said I didn't care, and stepped down into the grainy black mud, acutely aware that I was treading on a landscape of death and destruction, as well as sacred ground.

The two women got out on the other side. Manny and Elliot gave us the lay of the land, indicating where each building had stood. I knew I had to be by myself to home in on whatever it was I was supposed to learn by being here, so I walked away until I could no longer hear their conversation. Scott and Lou were leaning against the fire engine. Elliot and Manny and the ladies had walked some distance away overlooking the site.

I closed my eyes and opened myself up psychically. Screaming. Nonstop screaming. The two towers are outlined in screaming. My chest is crushed. I can't breathe. Massive pain shoots through my entire body, and I shudder, the vibrations taking over my body. My knees are weak and I nearly fall. I expected this, knowing what I know. It is horrible in its intensity, and yet I feel no profound change within myself. I had half hoped that maybe revisiting the trauma would heal the division in my brain, the constant ache, pull me back together again to where I was before 9/11. But there are no remarkable insights or feelings that would explain the compulsion to come to this place.

And then, in pieces, it comes.

I look over at Scott and Lou, and I see in the tired silhouettes what they will never, ever, show me in the firehouse. The twin towers are bearing down on them, crushing the life out of them. In their minds, and in their bodies, they have imagined countless times the death of their brothers. They knew these buildings; they had been to them – and in them – hundreds of times. Their eyes reflect all they had seen and been through, and it was not over yet. They could not get away from it; they were buried in it. Elliot and Manny were still talking to Linda and Abby in absolute calm and peace; the perfect hosts. Their graciousness was excruciating to witness in the terrible space where they stood. Who were these men? How could they do this? How could they stay so centered when they were enduring the worst grief and pain of their lives?

The only thing they had to offer all those who had come to their door was to bring them to the place that haunted them. To the graves of their brothers. How many people had come by? How many times had they revisited the aching emptiness of the site? How could anyone let them do this?

Something in me snapped. We had to get out of here; we had to get these men out of here now. I turned toward Linda and Abby, hoping to God they would take the hint and bring this to a close. I strode purposefully toward the fire engine to indicate it was time to go.

And at that moment, in a tired voice that burned into my psyche, Scott said: "I hate coming here. I hate this place. I can't wait for the day I never have to come here again."

Lou was silent, but his downcast eyes and bowed shoulders spoke volumes. There was nothing I could say. Nothing. I stood with them, noticing how black the night and the ground were, even though the white lights hammered us with their brilliance.

It seemed like an eternity before the others came back to the fire engine. It was a subdued ride back to Engine 32; I don't remember what was said, if anything. I was too overwhelmed with what I had seen and felt.

The women asked if they could take a picture of the firefighters in front of the engine after we had backed into the house. My mind was protesting: No, no, don't ask anything more of these men, no more.

Of course, the men graciously gathered for the picture. There was something about the way they did it that showed they had been through this a thousand times already. As Linda fumbled with the camera, Lou took it from her, saying, "I'll take the picture, you ladies get in there with those guys."

As the ladies protested, I spoke up to get it over with. "Let me take the picture, I'm not part of this."

Scott and Elliot immediately and emphatically said, "Yes, you are, get in here." So we stood for the photo. Nobody smiled. Then everyone said their farewells.

I asked Elliot if he wanted an Alexander Technique lesson. He protested, indicating that I must be tired as it was, after all, 10 p.m., but I said I wasn't, and I wasn't. What I had seen and experienced at Ground Zero that evening had triggered a massive release of energy within me. I knew what I had to do, why I had been called down here. I was to take care of them. Elliot shrugged, and we went upstairs.

He was a veteran, with perhaps 19 or 20 years on the job, and even though he was technically the "client," in 20 seconds flat he had turned the tables and was trying to be a therapist for me. I decided to let him run the session, knowing that the Alexander Technique would speak for itself in the end, no matter who did the talking. Honestly, I was not sure that these men needed me or my work. I now knew that they were highly-evolved and somewhat superhuman, and maybe they had the ability to get through this horror without outside help.

Elliot wanted to know how I had come to Engine 32. In my purse I carried the CD of my composition, which was finally titled: *Reading of the Names 9/11: The Firefighters.* I was simply waiting for the right time and place to present it. I opened up to him, trusting him completely, telling him the whole story: How I first came to New York, and how being at Engine 32 had changed something inside me, how my friends were still astonished that I had actually left my apartment, and about the check for $3000 miraculously showing up, paying for my hotel.

Elliot, concerned about my finances, said I could stay overnight at Engine 32. I said, "Isn't that against the rules?"

"There are no rules right now. We've had a lot of people staying here, you're helping us out, and you're welcome to stay." I thanked him for offering, knowing I would never do it.

Would he like to hear the CD? Yes, he would.

The music which had dominated my entire life and living space while coming into existence now seemed so small set against the magnitude of the men about whom it was written. It made its world premiere in the bunkroom of Engine 32. Elliot said it was nice. And that was it.

I called Lou in from the office to begin his lesson, but he protested that it was too late for me. Now, however, I was determined and invincible, and I refused to take no for an answer. I overheard him say to Elliot "What stamina she has!"

My friend Dawn would say that I needed to take a nap in order to gather enough energy to go to bed. Was he really talking about me? Here in this firehouse I was being redefined into something powerful. It was intoxicating. I wanted to do more and more. And on the other side, I still waited for them to realize I was broken. Eventually, they would uncover the truth. Eventually, I would do something stupid and give it all away.

Lou did competitive weightlifting, and he was huge. Trying to work with him on the pathetic beds, sagging in the middle, was a joke. He was tired, and drained, and I prayed that the energy I felt would translate through my hands and benefit him in some small way.

I left him there slumbering and hoped that no alarms would sound. Downstairs, the men were watching television. Scott greeted me and said, "Are you sure you don't want to stay here?" I assured him I was going to be fine at the hotel. He brought my luggage down, and walked me to the front of the firehouse. We hugged each other briefly in understanding of all we had been through. It had been a long day.

I leaned back against the seat of the cab in utter satisfaction. Once again, I was driven up Barclay Street and I asked the driver to slow down as we approached Ground Zero. And once again, all the pain and fear were gone. I was solid as a rock.

My hotel was on West 23rd Street. I had no idea of what to expect, since it was the cheapest place I could find – $70 a night. Walking in, my life went from color to black and white. The atmosphere of the firehouse was so warm, with its vivid reds and blues, and this was … so not. Somehow, at some point, I realized the place was run by nuns. There was a crucifix and a Bible in my tiny little room with its child-size twin bed. A small window faced a brick wall. There was something about going from all those muscle-bound and glorious men, the oversize rooms and 16-foot ceilings, to this tiny cloistered cubicle of austerity, that had me giggling nonstop.

I unpacked my clothes, hurriedly washed up, and crawled into bed, exhausted. But I was happy, and content. I had been at a firehouse for 12 hours, given four Alexander Technique lessons, ridden in a fire engine, stood at the center of Ground Zero, and worked until midnight. What stamina I had.

TUESDAY, FEBRUARY 26, 2002

My first attempt to get out of bed was just that – an attempt. My arms ached and my legs were stiff and sore from climbing the firehouse stairs and getting in and out of the fire engine. The very idea that I might have to repeat the arduous tasks of yesterday was overwhelming. But the memory of the 12-hour bucket lines pushed my tired body upright. There would be plenty of time to rest once I returned to Boston.

After showering and dressing, I reached for my black loafers. I winced. Dried mud, the sacred ground of the World Trade Center site had collected in the crevices around the edges and in the rippled soles. I didn't want to show up at the firehouse with mud on my shoes; but the idea of callously scraping it off into a waste basket or washing the mud off in the bathroom sink ... well, I simply couldn't do it.

I would have to collect it, somehow, and figure out what to do with it later.

Using a nail file, I scraped the mud from the shoes as best I could onto a tissue, and carefully folded it into a packet. I tucked it into one of the inner side pockets of my suitcase, in case anyone came in to clean the room.[1]

The firefighters had informed me that "BI" (building inspections)

---

[1] It now resides in an antique Blue Willow covered dish, resting on a bookcase in my apartment.

were carried out from late morning to early afternoon on Tuesdays, so it was best for me to arrive around 2:00. Still not confident enough to handle the subway system, I took a cab to the firehouse.

Manny met me at the door and was cordial and pleasant, but reserved. He escorted me past the fire engine and into the back dining area. There, four firefighters sat at the table, none of whom I recognized from the day before. Inwardly I grimaced, knowing I would have to start from scratch. How many guys were in a firehouse, anyway? Manny introduced me as a "massage therapist" and I could almost hear their minds clamping shut. Clearly, if the word 'therapy' was articulated again, my efforts would be doomed.

Thinking fast, I came up with a new sales pitch for the Alexander Technique. In my most convincing voice, I insisted that it wasn't massage therapy, it was a "new, state-of-the-art bodywork," more like a "tune-up." And then, "even if you are fine" it will "enhance your natural coordination," giving you "greater strength" and "ease of movement." In fact, if you had "no injuries" and there was "nothing wrong with you" you were more likely to have a better result from the lesson.

It worked.

Seeing me now as more of an amusement than a therapeutic threat, they cajoled and pushed the least resistant among them, Matt Dunn, into getting a session. I sensed that again they were going along with it only to reward my effort. Matt – sweet and shy – tried to protest, but relented to the pressure. I led him from the room, his head down and looking like a schoolboy on his way to the principal's office.

And so, this is how I met Matt Dunn, the sole survivor of the members of Engine 32 who went up into the towers on 9/11. I don't believe I ever understood the words "humble" or "selfless" until I met him. Yesterday's experiences with these firefighters had seriously undermined my beliefs about men, but Matt was about to further challenge my preconceptions. There was nothing harsh about him: He had a supple round face, and soft black hair. His initial response to any question was always to smile first; a sweet smile that would alter his shy demeanor. His blue eyes would sparkle, reflecting the existence of a loving spirit. He brought a gentleness to the conversation that made me feel safe. No matter what I said, it would be fine with him. He melted me. Any and all defenses I held within myself were unnecessary in his presence, and I hardly knew how to be around a man so unprotected by ego.

An alarm sounded during his session, and he asked if I wanted to come along. Of course I did. He held the door open for me, and noticing my difficulty managing the high step, said quietly and gently, "jump up, don't climb." I understood this immediately, and found myself navigating the height with ease, accompanied by his simple praise: "You got it."

There was a timelessness about him, an understated presence. All of the philosophies I had studied about simply being – about living in the present – were observed and practiced with amazing ease in this firehouse. How could this man take the time to teach me to jump into a fire engine? Why would he? And with such kindness?

And yet, at the sound of that alarm the sweet shyness disappeared and he was transformed into a being of strength, drive and determination. Nothing stood in the way of these firefighters when called to duty. I would know by the end of his session that he was handling stress that would transform most people into blithering idiots. He had lost 50% of his breathing capacity from the scarring in his lungs. He had gone up into the North Tower (the first tower hit) along with Sean Michaels, Elias Keane, James Gluntz and Robert McMannis, a covering (substitute) lieutenant.

Matt asked me if I had seen the Engine 32 Memorial website, as his account of what took place was recorded there. I told him I hadn't, and encouraged him to tell me his version of the events that day.

For history's sake, and with his permission, I have reprinted the website version – *Matt's Story* – here:

∗   ∗   ∗

My name is Matt Dunn and this is what happened to me on 09/11/01. I am a fireman at Engine 32 of the New York City Fire Department, which is located just a few blocks away from the World Trade Center. I was working a 24 hour shift September 10, to go off duty at 0900 hours on 9/11/01. It was a good tour. I was working with a great bunch of guys. We had a covering Lieutenant supervising the company, Lt. Robert McMannis [covering means substituting in another firehouse]. He seemed to fit right in with our company. It was a pretty quiet night tour [shift]. Not too many runs.

At approximately 0815 hours Engine 32 received an EMS run for a medical emergency at the Governor Smith housing project. Engine 32 responded with the men working this 6 x 9 tour [6 PM to 9 AM]: Sean Michaels, Jim Gluntz, John O'Brien, Lt. McMannis and myself. We were returning from this call, waiting at a stop sign at the corner of Avenue of the Finest and Rose Streets, when we heard a very loud explosion. I saw a police officer pointing towards the World Trade Center. We looked up and saw a large hole in the north tower of the WTC. It looked as if a bomb had exploded in the north wall. The hole looked to be on about the 80th floor. It was a big jagged hole maybe involving three stories. Many papers came flying out, and smoke started to pour out.

Lt. McMannis reached for the department radio and transmitted a 2nd alarm and a 10-60 signal for the World Trade Center. A 10-60 signal is a code for a disaster. We headed towards the scene. First, down Gold Street. Two firefighters had run out to the street with their bunker gear. It was Elias Keane and Joe Gilmore. They were both scheduled to work the following 9 x 6 tour [9AM to 6PM]. They had heard the explosion too, and knew there was a job. They jumped on the apparatus. We were all crowded in the rig. Elias was very excited. He said this was going to be some job.

When the engine arrived about a block away from the WTC complex there were already thousands of people in the streets, trying to get away. Thousands jammed Vesey and Church streets. The apparatus had to slowly push the people out of the way to get through. Like a plow. Slowly pushing the people out of the way so we could respond. We were one of the first units on scene. The North Tower had multiple floors on fire, and there were people jumping. It was hell. It was like a nightmare. That was one of my most dreaded fears of being a firefighter: to witness people jumping from a building. It looked as if a few were landing on the roof of 6 WTC. We had to gather our equipment quickly and run in for fear that we would be hit by people jumping from the tower. Elias, Sean, Jim, Lt. McMannis and myself approached the entrance. John O'Brien, the engine operator, hooked up to the hydrant and prepared to supply the standpipe system. We entered though the front doors of the North Tower.

The lobby was a disaster. Every window in the lobby appeared to be broken, with giant shards of glass hanging from the window frames. These windows were over two stories high. In the lobby, our officer checked in at the fire command station. It was a few moments before the officers figured out what tactics we would use (attack stair, evacuation stair). The rest of us were readying ourselves for fire duty. Buttoning up our turnout coats, checking the straps on our masks, turning our air cylinders on. We started to realize we would have to walk this one. The elevators weren't going to be usable.

Jim Gluntz said, "Are you guys ready to walk" eagerly. Sean and I both shook our heads and said this would be a tough one. As Lt. McMannis hurried over from the fire command station and notified us we were going to start climbing the "B" staircase, I heard a radio transmission that it was a confirmed plane crash on the 86th floor. The members followed the Lieutenant to the "B" staircase, which was located in the core of the building, pretty much in the center. The elevator system was destroyed. Hoistway doors were ajar. Some were missing completely. I saw an elevator car in the shaft, twisted.

We advanced to the stairway. There was a lot of rubble in the corridor in front of the elevators. I remembered I looked up at the ceiling to see if it was charred and had fallen down. The ceiling in the lobby was maybe three stories high. The ceiling looked intact. We had to climb over and around this pile of rubble, which may have been three feet high in the middle. I thought maybe the floor had been blown upwards from below. Later I was to find out that the rubble we climbed over was burnt bodies.

We started up the stairs of the north tower with our hose lines and tools. Many people were coming down the stairs while we were going up. The stairway was so congested. We were telling the people to stay to the right. One long line of people trying to get down the stairs. Water was flowing down the steps like a rapid river. When we reached approximately the 10th floor, the water stopped.

About the 14th floor I heard a handie talkie transmission, someone screamed "a plane, another plane" then a small rumble. Someone said on the handie talkie that another plane had hit the other tower. The building shook a little and the lights flickered. We continued to proceed up the stairs. One file of firemen going up. Many people coming down. Some of the people were wounded. Some burned. Some other people were carrying victims. These terrified people were encouraging us. They were patting us on the back. "The firemen are here," they shouted.

When I reached the 17th floor my company was taking a blow [rest]. Lt. McMannis told us to take a blow while he headed up a little more. As I entered the corridor on the floor I saw a vending machine, which had a large glass panel. It had a lot of bottles of water and juice. Some of the firemen were seeing if they had change in their pockets. I grabbed the ax and smashed the glass panel. It took three hard blows to break the double-paned glass.[1] We all laughed. Sean and I started handing out water to the other fireman on the floor and in the stairway. We passed bottles of water to the people coming down the stairs evacuating. We rested for a few minutes more there on the 17th floor, and then we started to proceed up the stairs once again. People coming down the stairs were also handing us water bottles to drink from. We started hearing rumors that the Pentagon was hit, the White House was hit, and the Sears Tower too. We were all sweating a lot. When I reached the 31st floor my company was waiting for me. The Lt. said let's take a blow here for a minute. Other companies were also taking a blow.

I remember a Lt. from Engine 10 arrived on the 31st floor, and said, "Come on guys, we got to make a push." We were very fatigued from carrying our equipment and lengths of two-and-a-half-inch hose lines. We decided to leave two roll-up hose lengths there, and use four men to carry two lengths of hose, alternating between them. The officer and two men, Jim Gluntz and Elias Keane, started up the stairs. Sean Michaels took one roll-up and I started shortly after them. We were trying to make the 44th floor since there was a report of an elevator there. About the 37th floor the building shook violently. I donned my face piece quickly. The building really shook, tossing us around. Someone shouted, "The south tower has just collapsed." A few seconds later, an order was given to evacuate the building. I called my officer on the handie talkie, and he said he was making his way down. There was a lot of radio communication on the handie talkies. So much traffic between companies. It was taking a long time to get down. It seemed that you would get a couple of steps

---

[1] After Matt reported the destruction of the vending machine during his debriefing with the WTC Task Force, the battalion chief conducting the interview said, "Oh, so you're the guy they're looking for."

and then the movement would stop. A few more and then it would stop again. I remember I thought that guys were going to start panicking soon and I was scared I might be crushed by the panic. But no one did panic. You just waited until you could start advancing down the stairs some more. There was no pushing. There was no yelling. At one point I moved to the inside on the staircase. Scores of firemen were filing in from every floor. Some were leaving their equipment. When I reached the lower floors I had contact with Jim Gluntz on the handie talkie. He was making his way down.

When I got to the lobby I had to climb over that pile of rubble again. There were a few inches of dust and debris that were layering the floor. The front doors by which we had entered the building were now blocked off by mounds of rubble. I saw that the fire command station was empty. I was with about 20 other firemen when we reached the northwest window. There was a man dressed in a dark blue uniform standing about 75 feet away from the building looking up. He was covered in dust. I couldn't tell if he was a fireman. Everything was covered in dust. He pointed up and held his hands out. We took that as a signal to wait because something was coming down. It was more jumpers. He waved us to start coming out. I had to climb out a window on West Street. I remembered I looked at this pile of bodies that was in front of the building. It looked like a large pile of dead black cows. I remember I looked to the left and saw that Three World Trade Center had a giant "V" section of the building missing. It looked like from the top floor to maybe the third floor was missing completely. There was a lot of rubble all around. I saw the Engine 32 Pumper with it lights revolving and no personnel around it.

I started north on West Street with many other firefighters, when One World Trade Center started to collapse. Someone said it's falling and everybody started to run. I looked up and saw what looked like the top 20 stories of the building starting to lean towards me. I tried to run but couldn't outrun the cloud. I tried not to look back and run for fear it would slow me down, but I kept looking back, I donned my face piece and got my helmet back on just as the blast came.

I was first enveloped in papers on fire with some heat. It was all orange around me. I could feel the heat on the back of my neck and ears. Then thick black smoke and dust. I was choking and spitting up in my face piece, from all the dust particles in my mask. I knew that any second something solid would hit me and I would be killed. I could not see at all. I was scared I would be killed and my wife would be left all alone. I was pelted with small particles of debris. I stopped running. I turned my flashlight on. I could see a slight beam shine out maybe 2 to 3 inches. I walked for a while in blindness with my hands stretched out in front of me. Stumbling, not being able to see anything. I walked a long time. It seemed very far. I was wondering if my air would run out.

I tried to control my breathing. I was coughing a lot in my mask. My eyes were all irritated. I was wondering if I would run out of air. Then all of a sudden the smoke started to lift, and I could start to see a little. I was tripping over things on the ground. I took off my face piece. It was hard to see and breathe. There was a news cameraman there, videotaping me. I remember I waved him off. Two police officers started to help me; they put me on my knees and tried to wash my eyes out. They helped me back up and escorted me towards the high school where a triage center had been set up. I entered the lobby of the school and there were hundreds of people. I felt I was all right, and wanted to regroup with my company. I headed towards West Street, where I ran into many members from my battalion. Some companies were missing members. I tried to contact my company by radio. There was no response. There were a lot of maydays being given over the radio. All of a sudden people were running north up West Street. I feared I was going to be crushed. I couldn't run. A wall of people was coming at me. I went over by the sidewalk and fell to my knees. I was so fatigued, I could hardly move. After a while an army paramedic came up to me and asked me if I was all right. I told him I was all right, just a little cold. Then I thought, it was a very warm day, why was I cold. He took my vitals; he thought I might be having a heart attack. He gave me a nitro pill. It made me feel like jelly. He found a couple of EMTs and they found an ambulance that had its windows blown out. They put me on a stretcher and put me in the ambulance. They took me to St. Vincent's Hospital. The hospital had a decon [decontamination] team outside. They took my bunker gear off, washed me down with sponges, and rushed me into the emergency room. There it seamed like a whole team of doctors and nurses were attending to me. Asking me a bunch of questions. Taking X-rays. Looking for injuries. After evaluating me, I was told that they didn't think I had a cardio episode, but I did inhale a lot of contaminants. They admitted me to the hospital. I remember they were pushing my bed into the elevator and it made me very uncomfortable to be there. I didn't want to be inside a building. I think they brought me to the 11th floor. They put me in a room and hooked up an I.V. A nurse asked me if she could contact anyone for me. I asked her to call the firehouse to let them know I was all right. I asked her to also call my wife. I was coughing a lot. They were giving me nebulizer treatments. A short while later the nurse told me she had contacted my firehouse and my wife. My wife came to the hospital and stayed with me the next 30 hours. It was about midnight when my captain and some members of my company found me in the hospital. My captain said they needed to know where we were in the building, because there was very little information about where we were operating. I described where we were and what staircase we were in. They didn't tell me, but I could tell in their eyes that no one else from my company had made it out. I remembered after they left I turned on the TV in my room and only one channel would come in. Channel 2, and it wasn't very clear. This is when I learned that two jetliners had crashed into the WTC.

Cardinal Egan came to the hospital to visit me. I asked him to pray for my missing brothers. They continued to give me treatments for my lungs and I was released on Thursday afternoon. Some police officers escorted me down to the front of the hospital. There were a lot of press people waiting with cameras. I asked the police if they could keep the press away and if they could bring me to my firehouse. When I arrived at the firehouse, I started to cry. The cars of Jim, Sean, Eli, & Lt. McMannis were still parked out front. I then learned for sure that they were all missing.

I felt so empty inside. All of the guys were saying they were so happy I was all right, that I had survived. They were all giving me hugs. I felt strange. Why couldn't the rest of the guys have made it out? There is no answer. I had survived and would live the rest of my life wondering how I made it. I felt bad. I missed the guys. I was told an extensive search was being conducted.

The firehouse was like a war zone. Every inch of the firehouse was covered in this gray, chalky dust. We had no fire engine. It was destroyed when building 1 collapsed. No tools, no equipment, no breathing apparatus. A lot of the remaining gear had been taken because teams of brothers were actively searching the wreckage pile. They said there were a lot of voids and they were conducting an intensive search. The air downtown was thick and dusty. We were all coughing a lot. There was some kind of FBI command post set up right in front of the firehouse. Many retired members had come to the house and were helping with search efforts. I remember about eight o'clock that evening that two retired members brought me home. I was reunited with my wife. We were so sad about what had happened. I spoke to Sean Michaels's wife that night on the telephone. We both were crying. That night I tried to lay down and sleep about midnight. Everything just kept playing over and over in my mind.

A friend came by in the morning and took me down to the firehouse. We teamed up with a few brothers and got some work gloves and they handed me a paper mask. The air was so thick and dusty. There was a burning odor in the air unlike anything I had ever smelled before. It was very irritating to my throat. We headed down to the WTC. Many buildings were damaged, even blocks away from the site. When we finally got over to West Street I could start to see the mass of the devastation. It was like a nightmare. This wasn't happening. Everything looked destroyed. I was shaking. Coughing a lot. The paper mask really didn't do much. I could see a little down West Street. I could see our Pumper was destroyed under the covered footbridge that used to connect the WTC with the World Financial Center. We walked through the World Financial Center complex, out on to Liberty Street towards West Street. There were many ironworkers removing large I-beams. There was a long line of men passing buckets with debris. We joined the line and kept making our way forward. Climbing the pile, passing wreckage down the line. We were on what had been Three WTC. I said to the men, we were searching the wrong area. Our guys were going to be over towards were Tower 1 once stood. It was a pile at least 10 stories high. I remembered I was upset. But I was assured they were working in that area earlier and the cranes had to pull some wreckage away for us to advance. I was coughing a lot. The air was so irritating and I was choking from it. After a few hours, we went back to the firehouse. I stayed at the firehouse the next couple of nights. I couldn't sleep. I was walking around like a zombie. I went outside and I walked down to the hospital. There were hundreds of paper posters with pictures and descriptions of missing people. It was such a nightmare.

*   *   *

I couldn't get it out of my mind. It wasn't what he said; it was the way he spoke, like he was preparing me for something that was yet to come. Someday I would draw on the memory of how he had handled this. Even more exceptional, how had he managed to tell me a story of such horror and yet manage to keep me from feeling it? My ability to psychically absorb and analyze my clients' emotional and physical pain had been an infallible resource in treating them holistically. But here, this would change. Before the week was over, I would find that every man who received the Alexander work would not release their emotional pain or suffering. Instead, I felt lifted up in working with them, almost as though they were giving the session back to me. How was this possible? People on the outside were telling me the FDNY had been brought to their knees by this tragedy, but my experience contradicted that premise. Were these men intentionally protecting me from their pain?

I didn't know.

*　　　*　　　*

From 5:00 to 6:00 the shift changed. Men were arriving in civilian dress, exchanging good-natured and affectionate greetings, then going upstairs to change into uniform. Ray was back. So was Harry. We were so happy to see each other. But Harry had been "detailed" (meaning he would be covering for someone else) over at Engine 7, so he would not be at Engine 32 that night. I expressed my disappointment, telling him a second lesson would have really provided a great benefit. He smiled from ear to ear, and told me he felt great, he was fine, I was not to worry, that I should take care of the other guys.

Elliot walked in the door and I sighed with relief, knowing that my chances of getting men to participate and let me work with them were better with an advocate on the sidelines. He asked me if I wanted to go out with them to get the meal. Today I eagerly accepted. I wanted to learn more about the lives these men were living. In my head I formed a backup plan. If they had to go to a fire, I would always be sure to have my purse with me so that they could just leave me anywhere and I could take a cab back to my hotel.

We got in the engine and headed out – to where, I don't know – but soon we were under the Brooklyn Bridge. It was amazing to see this major NYC landmark for the first time, especially since I was viewing it through the window of this beloved engine. We pulled into the parking lot of a Pathmark grocery store. There were two other fire engines already there.

I was very surprised. I felt so sorry that these heroic men were somehow reduced to something as mundane as food shopping. Elliot stayed in the rig, and the guys asked me if I wanted to come in with them. As we headed across the parking lot, I asked a lot of questions. You do this every night? Yes. You pay for your own meals? Yes. What happens if there's a fire while you're in the middle of shopping?

"We just leave the cart and go. The store is used to it."

We walked in through the automatic doors, Ray grabbed a cart, and we started down the aisles.

Again I felt like a child as I followed them around the store. In this vast city these men were all I knew, and instinctively I stayed close, so as not to be left behind. The energy coming from these firemen was quietly challenging me to surrender myself to their care. I could feel it happening, and I sensed myself wanting to give in to it, but I feared it would eventually lead to the sexual overtures of my past. I held back.

There were two other sets of firefighters from other houses in the store. They acknowledged the men from Engine 32, and looked at me curiously. I looked back with equal curiosity, as though we were on opposite sides of a fence at a zoo. I had never, to my knowledge, seen firefighters shopping in a grocery store for their meals, and if I had, I would have thought they were in the store because of an alarm.

I offered to pay for the meal.

"Absolutely not!"

We returned to the firehouse, the grocery bags were carried into the kitchen, and the men went to work. I asked if I could help: "Absolutely not!" You never saw such an operation in progress; three men chopping and shredding, the clanging of pans, washing of broccoli, scrubbing of potatoes, a big loaf of bread cut into pieces, chicken dipped in egg, rolled into breadcrumbs and splashed down into the oil popping on the stove.

The Engine 32 kitchen was a long, skinny room, and no matter where I stood, I was in the way. There was a long stainless steel counter with two deep sinks. A gigantic can opener was bolted to the counter. Directly across was a large white refrigerator that had seen better days. Stacks of huge, badly-dented aluminum pots were balanced on the top. A monstrous old gas stove dominated the narrow kitchen like a fat and grease-blackened queen on a throne. The pilot light in the far corner of the upper broiler radiated a comforting warmth in a gentle gold and blue flame.

The counter next to the stove consisted of cheap grey plywood cabinets, and the drawers went off their runners when pulled open. Along the back of the counter was a makeshift shelf which held large bottles of spices, sugar, tea, olive oil, and coffee. Two coffeemakers dominated the end of the counter, which was old and heavily stained. Ugly bricks covered the back wall; some were chipped, many were simply missing. Mounted to the wall was an old cast iron dinner bell, with a large number "32" embedded into the brick beside it.

Another beaten up refrigerator completed the appliance ensemble, and this one was stuffed solid. Tall cabinets stood next to the refrigerator, containing cookbooks and other miscellaneous kitchen items. A dartboard hung on the wall at the end of the kitchen. The last counter held a toaster, mixer, huge rolls of saran wrap and aluminum foil, piles of brown paper towels, and various baked goods sent from people across the country, many with notes of goodwill still attached.

And lastly, a most peculiar smell: Years of grease and diesel and Pine Sol and brewing coffee, with some dishwasher detergent thrown in, and a touch of ammonia from the blue spray bottle.

I found the farthest wall to lean on and watched. I felt so bad for them, having to cook their own meals, remembering the prison that making dinner had been for me in my childhood. And they had to do this every night, and work too. It was insane, and it was sad. But unlike me, they seemed to take everything in stride. They were all working together and talking shop, almost like they enjoyed it.

They were absolutely fascinating. Adopting this firehouse was providing me with an opportunity to finally study an entire group of men in their natural habitat, without tainting the study by being in a romantic relationship at the same time. My first observation, and the one which keyed me into why my relationships had failed: They had a different way of speaking. It wasn't the kind of conversation where people had to dominate. There was a spaciousness to it. They waited, and when it was their time to contribute, they did. It was simple and unforced. It was also efficient. Long silences were part of the conversation, and no one seemed uncomfortable leaving it there. Their language made sense, based on a forward linear framework. Action. Solution. So unlike the way women would go on about their feelings with no desire for resolution. My silence was not out of place here; instead I was certain it was my only place. They were aware of me, and I was given a certain amount of space to exist in, but unless there was a clear cue for me, I was not supposed to speak. And I didn't. At first I found it disconcerting to have so much empty space, but as the evening progressed it brought about a quieting to my being which was oddly pleasant.

There was a new guy sitting at the table that evening. His name was Dan, and Elliot introduced me as a massage therapist. I offered him a lesson, with the usual comments about it not being a massage, really. He said he might do it later, but that he had a lot of paperwork to complete. He was very tall, mustached, big-chested guy, with a full head of unruly brown hair. He sat across from me, and kept his eyes down. Ray sat across from me, too, and he didn't look at me either. There was a lot of silence all around. I felt bad that my presence was causing such timidity, but I couldn't imagine saying anything that wouldn't sound forced. So we all watched TV.

I got up to clear my plate, but somebody pushed me back down and grabbed the plate from my hand. "You're the guest, don't you move." I protested, but they ignored it. I asked Ray if he wanted a lesson; his only response was a curt, "If it ain't broke, don't fix it." The table was cleared, and Elliot and I went upstairs, having agreed the night before that I would work with him if time permitted. With Harry gone, and maybe only Dan wanting a lesson, it looked to be a slow evening and I would likely get back to the hotel early.

Elliot totally trusted me, and this time we didn't talk much. Alarms came in, I went on them all, and although I was thrilled to be going, I was more able to control my delight. I looked around at the men, and wondered if they were smiling inside, too. When we got back I asked Dan again if he was going to have a lesson, and once again he offered the paperwork excuse.

With all the interruptions, it was 10:30 before I finished with Elliot. Dan was a no-show, and I waited downstairs with Elliot and the other men for another half hour, watching TV. Finally, Elliot said, "Just go knock on the door and say, "Dan, it's time."

I winced, saying, "I don't want to push him." And Elliot again said, "just say, 'Dan, it's time."

I went upstairs and knocked on the office door. "Hi, Dan, are you ready now?" He yielded his paperwork. I worked with him in silence. My desire to do something for these men overrode my exhaustion. Afterwards, Elliot walked me out to get a cab. I rolled into my room at one in the morning, wondering what the nuns were thinking.

I had to check out of what I was now calling The Convent by 11 a.m. But for some reason it didn't feel right to be going home this evening. I called an old client and fellow musician, Leon, who lived on St. Mark's Place, and asked him if he could put me up for the night if I decided to stay. He was happy to, so I swung by to drop off my suitcase and get a set of keys. Seeing someone I knew here in this city was so emotionally difficult, and I found myself crying despite my best efforts.

I showed up at the firehouse around 1:30, and again Manny met me at the door. I said in amazement, "You're still here?!" He had worked three days straight and looked exhausted, with dark circles under his eyes. Three men sat at the dining room table, and I sensed that this was going to be a hard sell. Ray would not get a session. Manny had already declined two days in a row. The looks on the faces of the other two didn't give me much hope. One of them was John O'Brien; white-haired, blue eyed, easily in his 50's. You could see the stress of the job and the aftermath of 9/11 etched in his face. Physically, he was not a big man, but he stood ramrod straight and solid at all times. He, like Ray, had a harsh, impassive look that communicated no interest in the subject of bodywork. Being female would get me no extra points or credits. I was given no opening to compel them with my marketing pitch. My sense was that they were exhausted by people invading their firehouse, and they just wanted things to revert to some semblance of normalcy.

It was an awful moment, knowing that they didn't want me there, but when I tuned in the "voice" was telling me I should stay. The men offered me some lunch, but having already eaten, I declined. They then proceeded to ignore me. I sat in silence, wondering how long this was going to continue. Finally one of them, Steve, announced that he would take a lesson, and we went upstairs.

Steve didn't trust me or let me in. He was a very good-looking guy, but his appearance was flawed by a continuous frown on his face. He submitted to the lesson, but wouldn't open up to me or give me any credit for my work. I didn't take it personally. I had a bigger mission, and Elliot's and Harry's total acceptance of me was validation enough. However, I screwed up. In order to have something to talk with them about, I had started watching sports. At that time, the New England Patriots were winning the Superbowl. I had watched the post-season games, with Adam Vinatieri kicking the winning field goal in the snow, but for me, it was all a schmear. I talked with Steve about the games in an attempt to break down his wall, and he asked me if I remembered a specific play. I couldn't remember the play, much less the team, and I fumbled the stupid girl ball in front of him. I don't think he trusted me much after that.

I fail to understand how men can go through the shock of a lifetime, yet still remember who did what with a ball.

But we did have a good session, and that helped ease the faux pas. As we were finishing, we heard the downstairs door open, and I hoped it was a firefighter who would be a little more enthusiastic about my presence there. I asked Steve if he would send up whoever else might want to work with me.

Steve went through the door, and I heard him yell down the stairs. "Hey, Robby! Robby!"

A voice yelled back. "Yeah?"

"You want a massage?"

A very nervous, uncomfortable voice responded, "From you?"

Steve exploded angrily, "No, you fuckhead, there's a massage therapist upstairs!"

The relief from down below flooded the air. Laughing, it was all I could do to pull myself back together before he made it into the room.

Robby was a "detail" from Engine 7, covering for someone absent from Engine 32. He was very young and inexperienced, thinly built, with a very serious look. As I worked with him he felt extraordinary changes taking place, and he expressed his intrigue and amazement at the nature of the Alexander work. He had been working on 9/11, and he told me his job in the weeks thereafter was "bagging, tagging and flagging" body parts. I looked down at this young kid and wondered what effect this would have on his psyche years from now. He said, very matter-of-fact, he had been through worse. I was afraid to ask him what worse was.

John O'Brien, the white-haired veteran, came in the room during the treatment, and stood watching for awhile with a frown on his face, his fingers stuck in his belt. He grabbed a pillow and put it over Robby's face, pretending to suffocate him. "I guess I won't be getting much work out of you today." He left, his obvious disapproval of my presence lingering in the air.

Afterward, Robby asked if I was staying for dinner. Since none of the other men wanted a session, I told him that I should leave. He mentioned that John was cooking that evening so dinner would, in all probability, be Sloppy Joe sandwiches. I thanked him for the heads-up. I had eaten more bread, potatoes, red meat and cheese in the last three days than I had in the last 30 years. I had to draw the line at Sloppy Joes. I went downstairs and made a point to say good-bye to everyone, but unlike the past two days, there was no invitation to return.

John came into the dining area. With absolutely no expression on his face, he said, "You left your jewelry upstairs." I cringed. I had taken off my ring and bracelet in order to work, and had made a terrible mistake in forgetting them. This guy didn't want to see any woman's jewelry laying anywhere in his firehouse. I knew then that John didn't want me there. I didn't take it personally, although I was embarrassed to have put myself in such a position. Surely, all of the men were desperate to get back to some kind of normal, and my presence served as a reminder that it wasn't.

My desire to help was creating just the opposite, making the men more stressed. They were like wounded animals. If I couldn't work with John, the next best thing I could do for him was leave, and after making the humiliating climb to retrieve my jewelry, I did.

I made a decision: If the reception was similar tomorrow, I would have to, out of love, leave them alone for good.

I called my friend Leon to see if he wanted to have dinner, but he had a gig. I was on my own. He told me about a great little Italian restaurant called Col Legno near St. Mark's Place on the East Side, so I took a cab directly there.

The restaurant was small and simple, with a dozen tables and a wood-oven warming one corner of room. I was particularly pleased to see a cat wandering around beneath the tables, and felt very comfortable and at home. After ordering, I rested my chin in my hands and stopped to reflect on all that had transpired over the past few days … Matt's extraordinary selflessness in particular. Overcome by his sheer goodness, knowing what he was facing, realizing I, the so-called "healer" was not able to heal this, I found myself crying. Julio, my waiter, was sensitive enough to wait between rounds of tears to bring my food or take it away. He didn't know why I was crying, but the whole city was crying in one form or another. I was just one more.

THURSDAY, FEBRUARY 28, 2002

In a final attempt to save the project I called the firehouse and asked if they would contact Harry to see if he would come in early for a session. I don't know who I talked with, but they seemed happy that I was coming over, and I set out much relieved.

I was met at the door by a firefighter who was tremendously amused that an "Alexander Technique teacher" had come to his firehouse. He said he was definitely going to take advantage of it. He introduced himself as Andy, and in minutes we were laughing ourselves silly.

"Have you met Harry? You know, that guy never stops talking." He proceeded to tell me that he and the other firefighters had a plan to kidnap Harry's kids and sell them to Turkish slave traders, and that the kids would thank them when they grew up.

Andy shared ailments with Matt Dunn, notably, the loss of lung capacity due to the inhalation of fiberglass "and God knows what else." Andy told me they couldn't use masks that day at the site, they just got clogged up. He was now paying the price, coughing so hard he would throw up nearly every morning. I saw a flash of worry cross his face, even though he spoke matter-of-factly about it.

We finished, and went downstairs, and another firefighter came in the door. As it turned out, most of the men were out on building inspections again, so I was lucky that anyone was here at all. Andy talked the other firefighter, Tom Hogan, into working with me, and back upstairs I went.

He started to remove his shirt. "No, no, leave your shirt on, this is different, this is state-of-the-art bodywork. Trust me." He shrugged, frowning slightly, not sure of what he was getting into. But once I started, he immediately "got it." He was 26, bright, sharp, and blunt.

Tom told me 9/11 was his day to work, but he had asked someone to substitute for him, and that guy had died that morning. I'll never forget his tone of voice as he said it; guilt, grief and love woven in a single fragile breath. And then he pulled himself back. At this point, I stood hopelessly in awe at the level of honesty, openness and lack of pretense that existed in this firehouse, in these men. He went on to say he was glad I ended up here at E-32, because they hadn't received anywhere near the amount of attention or services that the bigger houses had after 9/11. I took note of this unsolicited confirmation that my presence here was still warranted.

He was amazed by the Alexander Technique. He couldn't believe the experience and the results. When he got up I said, "State-of-the art?" and held up my palm for a high-five. He high-fived me back, saying "State-of-the-art!" I asked him to talk it up a bit in front of the other guys, that some had been shy about trying it, and he said, "Sure."

Downstairs there were at least six firefighters in the dining area, four of whom I had already worked on. Somebody said, "How was it Tom, did you enjoy your massage?" Tom stretched his arms up over his head and said, "Man, I feel like a million bucks! That was the best massage I ever had in my life!" And he turned and winked at me. Andy was at the table, Richie was there, Harry, everyone seemed glad to see me. They asked how the nuns were doing.

Matt Dunn was in his dress uniform for his interview with the BBC. This was his second interview in as many days. I looked across the room at him and was profoundly moved that he was making this effort to preserve the history of this time despite the stress of doing so. The humility was so intense that I found it hard to be in its presence; could anyone be so giving? Eventually I would come to learn that Engine 32 had made major efforts to take part in the documentation of this tragedy, opening up to filmmakers, writers and historians to guarantee an accurate record would be made and preserved.

Andy asked for a business card. He said he traveled up to Boston every now and then. I told him that if he made it up to my area I would be happy to see him at no charge. I reached in my purse. Harry teased me about not giving him one sooner, and then, quietly, out of the corner, Matt said, "Can I have one, too?"

I stood there shining in the reflection of their kind acknowledgement. There was a real and solid connection with these firefighters that went beyond the sessions; something deeper that I could not explain to myself. Yes, there were holdouts, but I just needed to make adjustments, call ahead, and be sure the men who were working wanted to work with me. I renewed my commitment to come and visit them again soon.

We said our farewells, and Matt spoke a quiet, "thank you, Jessica." I walked away from the firehouse, suitcase in tow. There was an intense golden warmth across my back, as though their farewells had reached out and followed me. It was so strong that I looked back over my shoulder at the empty street.

*   *   *

With my head held high, I walked through Penn Station. Once more the fear of death was absent, and within me, there was much room left for the unqualified goodness of the universe to flow in. Like a dam bursting, the water of life poured into me and I relished in the sensation of utter happiness. Every movement had meaning and value. Clearly, Engine 32 made me strong, and the longer I was in that supportive environment, the stronger I became. For the first time, I heard myself say: "Life is good."

Life was good and now I, Jessica Locke, was about to buy a bowl of soup in New York City.

I counted up my cash: $20 for the cab to get home from the Boston train station, and $2.00 for some dinner. For $1.89 I ordered cream of broccoli soup to eat on the train. Everything I was doing seemed so special, every moment was precious. This was going to be the best bowl of soup I ever ate. I was so happy. I couldn't stop smiling, and as I turned to leave the counter, a young black girl standing near the door approached me. "You got any spare change?"

In that moment, I would have given her the world. But I had next to nothing. Sensing my hesitation was due to other reasons, she said, "I'm really just hungry. If you can buy me something to eat, that would be okay with me."

I had at least fifty cents in change in my purse, but knowing she was hungry impelled me to hand her my soup. "Here, take it. It's cream of broccoli."

She made a face. "Oh God, I can't eat broccoli." And she handed the soup back to me.

I was stunned. "Pardon?"

"My momma made me eat broccoli once, and it made me sick. I can't stand broccoli."

In my mind, I was thinking, "beggars can't be choosers ... *can they*?"

But hell, this is New York, the greatest city in the world, and even the beggars were confident. Even the destitute knew what they wanted. It was all so wonderful. I dug out every bit of change I had and gave it to her.

BOSTON, MASSACHUSETTS

A few days after returning home, I emailed Engine 32 to thank them for their hospitality. I also sent letters to Harry and Elliot privately expressing my appreciation for their kindness and, in particular, the glorious rides in the fire engine.

I never thought I would hear from any of them. So, I was surprised and touched to get the following email[1] from Dan, the shy and quiet firefighter I had to drag from his paperwork in order to give him a session.

From: Dan Fullerton <danfullerton@yahoo.com>
Date: Tue, 12 Mar 2002 10:18:30 -0800 (PST)
To: Jessica Locke <Jessical@shore.net>

Hi Jessica,

It is good to hear from you. So many people have done so much to help us with donations and kind words. You, however, have gone that extra mile. You have taken time from your own life and really been "hands on." It's a great comfort to us to know there are special people like you who can give of themselves. I've only been assigned to Engine 32 since December, but in that time I have

---

[1]This, and all other emails and letters, have been printed in their original format.

become very protective of these men. They have been through a lot and still have difficult times ahead in dealing with what they have been through and in supporting and comforting the families of the brothers that were lost. Your kind words and soothing hands have been a great help. Thank You.

DF

I thought it was very sweet that this new guy felt so protective towards the others; that he had taken it upon himself to care for them. And so I wrote:

From: Jessica Locke <jessical@shore.net>
Date: Tue, 12 Mar 2002 21:17:04 -0400
To: Dan Fullerton <danfullerton@yahoo.com>

Hi Dan,

It was so great to see your name on my incoming mail today! Thank you for that beautiful letter. It meant a great deal to me. I have been so deeply touched by the experiences at Engine 32. You guys gave back so much more than I gave. Know that my thoughts and love are with you all. I LOVE NEW YORK!!! I plan to live there someday.

jessica

This time I heard nothing back.

The atmosphere in the country reminded me of an enraged bear surrounded by stinging bees, flailing about to no effect. Bin Laden remained at large, and terrorist threats – real or imagined – became daily fare. A reporter from the Wall Street Journal had been taken hostage and brutally murdered, his beheading videotaped and posted on the Internet for the world to see and fear. These combined elements should have had me terrified and cowering inside my apartment, but it was not so. My certainty in the goodness of the men of Engine 32 neutralized the barrage of bad news that dominated the media, and confirmed my determination that the only way to fight this insanity was to continue to do good in the face of it … to "be the change you want to see in the world," as Gandhi said.

Still, I was in shock. My still-divided brain would not function at full capacity. I was forever forgetting where my car was parked. It was unnerving to suddenly lose bits and pieces of myself, yet I was consoled by the fact that the firefighters had been through much worse. None complained. They had become my role models for how to act, and with a childlike hero worship I imitated them, refusing to dwell on or talk about my symptoms or fears.

The cleanup of Ground Zero continued. I learned from Harry that firefighters were volunteering to be at the site a month at a time. I withheld my private concerns about the quality of the air they were breathing, while the Environmental Protection Agency reported that asbestos, mercury, lead, benzene, and chromium levels were below harmful levels. I didn't believe it. All you had to do was inhale. When you did, the air burned the inside of your nose.

Although Harry did not reveal the reason for the 30-day shifts, I knew that these men were struggling to bring dignity and respect to the recovery of victims' remains. Who would want their loved ones hauled off to a dump called Fresh Kills? Now that I personally knew these men, I was stupefied and angry that Mayor Giuliani would not allow them the time needed to perform the recovery properly but, instead, imposed a time limit. And then the incoming mayor, Michael Bloomberg, proceeded to close six firehouses for budgetary reasons, including firehouses that had lost men. Bloomberg's edict was a monstrous betrayal by the city that had been protected and defended by the firefighters. The mayor's lack of moral integrity in this regard – his inability to "do the right thing" – had many up in arms. With money pouring in from across the United States and, indeed, from around the world, not a single firehouse should have closed. But the city was unable – perhaps more so unwilling – to seek alternate solutions to the firehouse closings.

*   *   *

On March 11, 2002 a documentary titled *9/11*, filmed by Gideon and Jules Naudet, aired on national television. Although their original intent was to chronicle a young firefighter's first year on the job, the two brothers instead found themselves recording the devastation of that day. Indeed, their cameras were rolling to famously (and fatefully) capture the only footage of the first plane flying into Tower 2. We are then taken inside the lobby of the World Trade Center to witness the day's chronology from the perspective of the firefighters. Matt was kind enough to warn me about the sound made when those who jumped from the burning buildings landed on the plaza; a sound that once heard, no one could ever forget.

Determined to study every aspect of 9/11, I watched this film six times. The more I understood what the firefighters had been through, the better prepared I would be if and when they wanted to unravel their experiences.

BOSTON
MARCH 27, 2002

I was desperate to get back to New York, but unable to afford a hotel. Mentioning this to one of my clients, I was immediately offered a place to stay. Sue worked in Human Resources for a division of Marsh & McClellan, a company that had occupied the top floors of Tower 1. Almost three hundred employees had lost their lives.

"I have a furnished condo in NYC that's unused most of the time. You can use it if you want. Just let me know the dates you're going to be there."

And she handed me the keys. I was speechless.

With this incredibly gracious gesture the plan to adopt Engine 32 was now an official "go." All I needed was train fare, a little food, money for cabs, and I was on my way. I set the dates and emailed Dan.

From: Jessica Locke <jessical@shore.net>
Date: Thu, 28 Mar 2002 21:42:10 -0400
To: Dan Fullerton <danfullerton@yahoo.com>
Subject: Scheduling

Hi Dan!

I just wanted to let you know that I can be back in NYC

April 8-12. Is this a good time? Do you all want me back?

I don't know if I'm going to be able to get hold of Harry til after April 1, he said he was working at the WTC site for the month of March, (Been a rough month for all of you, I'm sure). Were any of the guys from Engine 32 found? I'm not getting much information up here on the news.

I look forward to seeing you and the rest of the guys, please give them all my warmest regards.

Take care,

Jessica

From: Dan Fullerton <danfullerton@yahoo.com>
Date: Fri, 29 Mar 2002 12:07 PM
To: Jessica Locke <jessical@shore.net>
Subject: Re: Scheduling

Hi Jessica,
It's good to hear from you. I hope everything is good up your way.

I haven't spoken to Harry in a while. He should be back in the firehouse sometime after April 2 and I'll let him know I heard from you.

One of the members who was lost has been identified. I just got a call about this last night and don't have many details. Any specific information regarding FDNY members is kept strictly confidential. His wife has been notified and has to make some difficult decisions. She has been waiting all this time to know something before having any kind of memorial. She will need some time and more information before she decides what she wants to do. I don't know her very well, we've met only briefly once or twice (I was only assigned here in December). I'm sure the coming weeks will be an emotionally difficult and hectic time for us.

I'm sure the guys would like to see you again. In the next few weeks though I'm sure they will be pretty much preoccupied in dealing with this development. We'll keep in touch and let you know what's going on before you make any plans.

Stay Well,
DF

His response hit me pretty hard. I knew I needed to be sensitive to what they were going through. But the longer my trip was delayed, the more likely they were to close ranks and close doors for good.

Who was this Dan, anyway? Why was some new guy making decisions for all the men? Didn't he realize that in a year nobody would care about what happened to them? Those who swore to "never forget" would forget. The shock would wear off, people would move on to the next crisis. The firefighters needed ongoing support in order to heal. I had to lay a foundation of trust; I had to keep showing up.

A greater fear kicked in. Was he on to me? He – maybe all of them – had figured out I wasn't who I had claimed to be, this strong person who knew what she was doing. I had been trying to convince them they needed this bodywork, when I, in fact, was the one who needed them.

And, like all the other good men in my life who I needed, they were going to put as much distance between me and them as possible.

He was just trying to let me down gently. They always seemed to know about the broken side. I had to let them go.

From: Jessica Locke <jessical@shore.net>
Date: Sat, 30 Mar 2002 13:48:45 -0400
To: Dan Fullerton <danfullerton@yahoo.com>

Dear Dan,

Thank you for letting me know what's going on. I've been very sad since hearing the news. My heart goes out to all of you at this difficult time.

I know the strength you find in each other will carry you through this, as it has with so much else in the past six months.

Let me know when you feel the time is right for me to return. I will wait to hear from you.

Take good care of yourselves. I know no finer men.
Jessica

MANHATTAN
TUESDAY, APRIL 9, 2002

Despite it all, I decided to go down to New York. The Memorial Lights, two intense sets of spotlights that would project an ethereal image of the Twin Towers into the nighttime sky, were going to be shining until April 13th. As a necessary step in my own healing around 9/11, as well as to do more research for my *Reading of the Names* project, I knew that I had to see this up close and personal.

Sue's condo was on the eighteenth floor of a hotel in midtown Manhattan. The location was unbelievably convenient, just 15 blocks from Penn Station. There was a 24-hour concierge and a doorman. The lobby was spotless, with gleaming gold and white marble floors. The unit had a decent-sized living room and bedroom. The little kitchenette had a stove, microwave and dishwasher. Cable, laundry, and a maid service to change sheets and clean the apartment were luxe touches. There was a view of Broadway, and if you leaned way out the window you could see a small slice of the Hudson River. The place was wonderful, and I said a prayer of thanks to Sue and to whoever else was watching out for me. It was clear that something – someone – wanted the men at Engine 32 taken care of. Now, if they would just let me.

I dropped off my luggage and immediately left again. I took the subway downtown for the first time. I asked a lot of questions, scrutinized the subway map as each stop went by, and managed to reach my destination.

Amazingly, I never once thought about being pushed onto the tracks to an early death.

Emerging on Fulton Street, I recalled that one direction would take me to the firehouse, the other to the World Trade Center site. My soul wanted to go to the firehouse, but instead, I forced myself toward Ground Zero. As I approached St. Paul's Church, I noticed the line for the viewing platform was almost non-existent. I got into the quickly-moving queue. I reached the guard, apologetically held out my empty hands as if to say, "I don't have a ticket," and was waved right on through.

The viewing platform was a long wooden ramp that ran the entire length of Fulton Street beside the church. It was simply built, with rough wood planking, evoking in my mind the crucifixion of Jesus or a platform for a hanging. I felt a sense of dread as I took the walk. In my head I heard Officer Weber's words from my first visit: *You don't want to go up there.* There was an eight-foot-high wall separating those going up the ramp from those coming down; once you got on this ride you couldn't get off. The ramp slowly ascended and widened as one walked to the viewing area at the top.

Strangely, the dramatic panorama revealed a sheer and devastating nothingness. I kept looking for something that wasn't there; buildings, people, images of that terrifying day that now existed only as an ethereal vision. In that moment, a strong current of grief overcame me. I thought myself utterly helpless, a healer unable to heal, only able to stand there as witness once again. I said quietly, over and over, "I am so sorry. I am so sorry."

I stayed for a couple of hours. There now stood a huge, orange manmade bridge which angled into the deep black pit; the old ramp area was being excavated. At the far side of the site – the corner of Liberty and West Street – I noticed the flashing lights of arriving ambulances and a fire engine. Men stood at attention while a body draped with an American flag was carried out on a stretcher. I recalled Matt telling me they believed many of the firefighters would be found buried beneath the dirt ramp that had originally been used to enter Ground Zero, as it was at the base where the South Tower once stood.

Night fell. The Memorial Lights shone, but there was a light fog stunting their ascent skyward. Sadly, they seemed more like two arms reaching for something they would never have. I stood there in support for as long as I could before the cold set into my bones. As I started up Fulton Street, I bypassed the subway as though it did not exist and headed toward Engine 32.

The streets and sidewalks were deserted. I crossed to the opposite side of the street from the firehouse and slowly approached. The lights were on, the engine was in; they were safe at home. At this time, they would be inside making the meal, tucked away in the little kitchen in the back.

My movements were restricted by Dan's words. I dared not cross closer to the firehouse. Yet, I found myself overwhelmed by a sad longing to be with them. Never had I dared to want anything before now, and here the cruel irony was that I now wanted something I could never have. The realization brought a sharp pain into my chest. What was wrong with me? I was 50 years old and was having feelings about a bunch of men? So much so that I was hanging outside their firehouse like a stray cat?

It was night. I was alone on a deserted street in New York City. It wasn't safe. I had to get out of there. I hurried up the street and took a cab back to the condo.

From: Jessica Locke <jessical@shore.net>
Date: Sat, 13 Apr 2002 00:39:48 -0400
To: Dan Fullerton <danfullerton@yahoo.com>

Hi Dan!

How are you? I was down in NYC Tuesday through Thursday, I decided to come down anyway to do more research, and to see the Memorial Lights before they turned them off. I don't know if you know part of the reason all this started was that I was trying to compose a memorial piece of music for 9/11.
Hard to write about something you can't comprehend.

I hope you and the guys are doing okay. I sent you all some good vibes while I was in the vicinity. I'm learning the subway's A-C-E line. Have you ever been on Boston's subway? It's like 2nd grade math compared to New York's!

Take care,

Jessica

From: Dan Fullerton <danfullerton@yahoo.com>
Date: Mon, 15 Apr 2002 08:54:46 -0700 (PDT)
To: Jessica Locke <jessical@shore.net>
Subject: update

Hi Jessica,

If you come down to the city again give us a call, if
things aren't too hectic we could have you over for
lunch.

A few of us will be up near Boston this weekend. The
city of Peabody invited some of us to their spring
concert: "A Tribute to the Big Apple". A few of us
are going with our wives and girlfriends. I have no
idea what kind of music it will be. Maybe you are
familiar with this event.

We have gotten invitations to all sorts of events around
the country. To speak at High School graduations,
march in parades etc. We can't make all of them but
we try to send at least one or two guys if we can.

People are still coming from around the country and
showing their support. I thought by this time things
would be getting back to "normal" but it looks like it
will continue through the summer. It has made me
realize how much this has affected the entire country.
I hope that people continue to come together. That
would be a great thing to come from all this, but at
what a cost.

A number of people have written and recorded songs about September 11 and sent us CDs. Most of what I've heard is Country Western and Gospel music. Some are better than others. I can tell a lot of emotion goes into these things. I hope your effort is something that is helpful to yourself and those who listen to it. It must be difficult; so many people have trouble even talking about it.

The recovery at the site should be completed in about 4-6 weeks. This will be an emotionally difficult time for a lot of people here. There are so many victims who will never be found. I am going to try to be there on the last day. There still hasn't been any decision made regarding a funeral for the member who was recently identified but I'll keep you informed.

This was supposed to be a short note just to keep you up to date but now I think I'm starting to ramble so I'll sign off before this turns into an epic.

Be Well,
DF

I was amazed by his letter; by the length of it, by the amount of information it contained, and more so that he was, as promised, keeping me informed about what was going on. My hopes were renewed that there was still a chance he might let me return.

From: Jessica Locke <jessical@shore.net>
Date: Thu, 18 Apr 2002 20:57:41 -0400
To: Dan Fullerton <danfullerton@yahoo.com>
Subject: Re: update

Dear Dan,

It is so kind of you to let me know what's going on. I really
appreciate it. I'm sorry I didn't respond right away to your
email, but I've been having a few rough days. I've been so
numb since 9/11 that I got used to it, thinking that this
was now the real me. But in the past few days, I've started
to have some intense feelings about what happened. So
much of it is just utterly incomprehensible. What must you be
going through?

You know, of all the guys I worked with the last time, you
are the only one that I know nothing about. Have you
always been a firefighter? Are you married? Kids?

Ah, Peabody. It won't be as exciting as New York,
but then, nothing is. I hope you all have a great time.
If you make it into Boston, my door is always open
to you.

Take good care,

Jessica

From: Dan Fullerton <danfullerton@yahoo.com>
Date: Fri, 19 Apr 2002 10:24:56 -0700 (PDT)
To: Jessica Locke <jessical@shore.net>
Subject: Re: update

Jessica,

I know how you feel. I get hit with waves of strong emotions from time to time. It can be very disconcerting. From what I've read and "experts" I've spoken to we are all experiencing Post Traumatic Stress Disorder to some degree. Many of us here have similar signs: difficulty sleeping, nightmares, anxiety, depression, intrusive images, even flashbacks. Some things that have been suggested, and that I find helpful are doing normal things, getting back into a routine. It's also important to take care of yourself physically: eat well, get enough sleep, exercise. The most important thing I've found is to talk about how you feel. An understanding friend or relative can be as effective as a professional counselor. I've told firefighters the best counseling will be talking to each other in the firehouse. Hang in there, you are stronger than you realize.

You asked about my situation so here goes. I was appointed to the Fire Department in 1979, I was 22. I was married for almost ten years but have been separated for almost nine. I have a son who will be graduating from high school in June. He's a great kid but doesn't know what he wants to do in life yet.

He seems to be handling things well but I try to keep a close eye on him. I have a wonderful girlfriend. You can imagine what she went through on 9/11, especially since I wasn't able to get in touch with her until almost 10 PM. She has helped me through this like no one else could. I'm very lucky to have her. When she says she understands she REALLY understands. I try to use her insights to help anybody I talk to who is having a difficult time.

I'll let you know how things go in Peabody.

Stay well and take care of yourself. If we all stand together we will be fine.

DF

From: Jessica Locke <jessical@shore.net
Date: Wed, 24 Apr 2002 01:00:35 -0400
To: Dan Fullerton <danfullerton@yahoo.com
Subject: a little history

Hi Dan,

I am blown away that you would take the time to address my issues concerning 9/11. The only thing I had noticed prior to this was the short-term memory loss for about 4 months after 9/11, and a sickening feeling in my stomach everytime I saw or heard an airplane. (Which is still with me.)

I don't know if you heard my history, so many guys asked me that I've kinda lost track of who I told what to. (good English--!) I was married in 1980, separated in 1983, and divorced in 1985. No children. Two cats. I've been in a few relationships since then, the last one ended 5 years ago. Now I just work all the time. No time to meet anyone.

In 1983 I was in a car accident leaving me with chronic back pain. The doctors said I would have to live with it. For three years I tried every form of body work that existed, sometimes getting three massages a week. Finally in 1986 someone directed me to this Alexander Technique teacher. He only saw me once a week for 45 minutes, but each session was amazing. The pain would come back, but it gradually diminished until in 3 months it had disappeared completely, and in 6 months all my migraine headaches were gone, TMJ pain, my posture and breathing had dramatically improved.

This seemed like a good thing to learn how to do. The rest is history ...

I'm also a video/film composer, I've done a lot of stuff for PBS. I love doing the music, but it's very hard to stay financially stable unless you are well-known.

Well, you take good care of yourself.
Warmest regards,

Jessica

From: Dan Fullerton <danfullerton@yahoo.com>
Date: Fri. 26 Apr 2002 08:15:05 -0700 (PDT)
To: Jessica Locke <jessical@shore.net>
Subject: Re: <no subject>

Hi,

It's good to hear you're doing better. The memory
loss is common also, I forgot about that (guess I had
it). The planes don't bother me too much anymore; I
live near LaGuardia airport so I've gotten used to
them.

Divorce, car accidents--life is hard, but that is what
makes us strong. It's good to see someone who was
injured and didn't just sit around complaining,
expecting people to feel sorry for them. You
sought out help and got better. Good for you.
This Alexander Technique is interesting (it sure feels
good). I'll have to learn more about it.

Our trip to Peabody was great. We had a great time.
It was good to get away, even for a day. The people
treated us like royalty. We were invited to their
spring concert: "A Tribute to the Big Apple". They
put on a great show. The town picked us up from
the station, put us up in a hotel and gave us a limo
to the concert. We stopped at the Firehouse and
met some of their guys. Then we were escorted
across the street to the town hall for the concert.
The Peabody Police stopped traffic and saluted

us as we entered the auditorium. We all sat
with the Mayor at his table. During the intermission
they brought us on stage and presented us with
donations from the Peabody FD and the town. We're
not used to this kind of attention but it was fun.

Take Care,
DF

From: Jessica Locke <jessical@shore.net
Date: Sun, 28 Apr 2002 22:50:52 -0400
To: Dan Fullerton <danfullerton@yahoo.com

Hi, Dan!

I loved hearing about your visit to Peabody, that
was so heartwarming! I'm so glad you guys got a
chance to get away and be treated so well.

I got some great news. A client of mine has a
furnished condo in NYC in midtown Manhattan.
She gave me the keys. She only stays there occasionally
on business, so I just have to coordinate with her
when I want to come down. I'll have a kitchen!
A real home away from home.

I would very much like to talk with you in person
sometime, are you at the firehouse next week? I
want to talk to you face to face about the men and
my concerns. I care about all of you so much.
And it's too hard to attempt through email. I can

have the condo May 6, 7, 8 are you around any
of those days?

Take good care of yourself, okay? My warmest
warmest regards to all the men,

Jessica

From: Dan Fullerton <danfullerton@yahoo.com>
Date: Wed, 1 May 2002 10:03:05 -0700 (PDT)
To: Jessica Locke <jessical@shore.net>
Subject: Re: <no subject>

Hi,

Sounds like you're doing better.

I'll probably be working Mon. night and Tue. day.
If you plan to come down give me a call Fri.
or Sat. at the Firehouse. Our schedule rotates
and sometimes we swap shifts but I can let you
know for certain then. Harry should be around
at that time also.

The guys seem to be doing OK. We're keeping busy in
the firehouse getting ready for annual inspection of
quarters. There are still other events we are
attending (there was just a street dedication for one
of the members who was lost).

The recovery operation at the WTC site is almost finished. There will be some kind of memorial service then. Exactly what it will entail hasn't been decided yet. It may be for firefighters and family members only.

Take care.
Talk to you soon.
DF

Yes. He had said yes! He actually wrote to me like I was – well – like I was somebody.

Finally, I gathered the courage to call the firehouse. Harry answered the phone. He asked excitedly, "When you coming down? Are you in town now? Where are you, are you here?"

My entire being breathed a sigh of relief. I told him that I had been trying to get down there for two months, but Dan had put me off and told me there was too much going on ... that it was "hectic."

Harry sounded puzzled and said, "It's always hectic. It's a firehouse."

I responded that Dan was probably trying to protect them from outsiders, but I remained frustrated that he did not see the benefit of what I was trying to do.

Harry said, "Well, he is the captain ..."

I didn't hear the end of the sentence. Dan was a captain? They had captains? Oh Lord. I frantically struggled to remember if I had said anything stupid in my emails. Oh Lord.

"Let me ask him if it's okay for you to come down, he's right here ..." Harry buried the phone while he spoke so I couldn't hear.

"Dan says you are a very nice lady, and it's fine for you to work at the firehouse. Did you want to talk to him?"

I went shy. "Oh, I don't know ..." but Harry put him on anyway. I

103

was barely breathing so afraid was I of saying the wrong thing and blowing the gig. I was relieved to notice he seemed as shy as me. We verified the dates, and I reiterated that I wanted to talk to him about the men when I got down there.

Okay. We were back on track. I would return to Engine 32. I couldn't keep the smile from my face.

BOSTON
FRIDAY, MAY 3, 2002
FILENE'S DEPARTMENT STORE

I stood in front of large glass cases filled with hundreds of bottles of perfume, feeling a little intimidated. I had forgotten how to be a girl.

"I'm looking for a perfume that men will like," I finally said to the clerk.

She didn't hesitate as she grabbed a bottle and made a motion for me to give her my hand. "Hanae Mori – it's sophisticated, complicated ... very elegant." She sprayed it on my wrist. I had to walk around the store aimlessly smelling myself every 30 seconds before I could commit. I decided to trust her.

Since my last visit to New York I had gone out for a daily walk to build more stamina. There was a low stone wall on my path that was the same height as the step into the fire engine. Each day, until I had it just right, I would stop and jump onto the wall, just as Matt had instructed. People driving by probably thought I was crazy, but I didn't care.

I bought some new clothes, having lost weight from the exercise – mostly in blues that matched my eyes. I had been wearing drab and frumpy baggy grays and blacks for so long. I replaced my sensible heavy duty loafers with a pair of hemp sandals that were so feminine, yet still practical enough for getting in and out of fire engines and running up and down stairs. I looked in the mirror and thought I didn't look so bad for 50.

MANHATTAN
MONDAY MAY 6, 2002

I called the firehouse to let them know I would arrive at 10:30 in the morning. This time I would start early and leave early. I wanted to impress upon Captain Dan Fullerton that I was a serious and competent professional. Being at the firehouse so late into the night wasn't appropriate for a lady. I could do better, and I would do better.

It was insanely hot for the beginning of May; temperatures were in the 90's and the subway was an oppressive oven. I exited at Fulton Street and walked up to street level. I was continually amazed at how lost I could be in this city. Often I would walk a block or more before realizing I was going in the wrong direction. This was one of those days; I walked toward Ground Zero instead of the firehouse.

On the left was a little café with a sign in the window: *Fresh Squeezed Juices.* Excellent. What better way to counteract the meat and potatoes diet of the firehouse than to consume a nice carrot, celery, beet, and parsley juice? How could you not love New York? It had everything.

I ordered a large drink, unwrapped my straw and, walking outside the door was astonished to see Engine 32 pulling up to the curb. My insides turned to mush at the sight of the men; it was so wonderful to see them again. Elliot was driving. He looked down at me and said, "Wow, you sure look good!" That made me so happy.

They had stopped to grab some cold drinks to guard against the heat of the day. Mark asked how the juice was and if the café was clean. "Seems to be," I replied, and he and two other firefighters went in to order. Elliot told me to get in the front seat, but I thought he was joking, so I stayed on the sidewalk and continued to exchange pleasantries. Dan came around to the front of the engine and we shook hands. He asked how my trip had been, and then walked into the café to join the others. Elliot and I talked until an alarm came in, then he radioed everyone to return to the rig.

It was an odd moment. They knew I was here to see them, and that I had no place to go while they responded to the alarm. (And of course, I was dying to ride in the fire engine again.) And yet, on a professional level, should I pretend I had other things to do? Wander through lower Manhattan in this god-awful heat until they returned to the firehouse? That's what a professional would do. I would do it. If they asked me to go with them, I would tell them we would meet back at the firehouse instead.

The men returned to the rig, and as Mark opened the door to get in, he stopped and looked at me. Very quietly he said, "Do you want to come with us?"

I crumbled. "Yes."

"Get in, but I have to make room for you. The mail is in the way."

I quickly scrambled in, not wanting to hold them up. But … there was no seat. It had been folded up to make room for the mail, which was in a large bin, but it looked as though the seat was simply gone. I didn't want Mark to concern himself with me, but I needed to get settled before the engine got going. So I sat on the floor.

Mark gave me an odd look, and then called up to the front, "Hey, Cap, Jessica's with us."

Dan looked back, and, unable to see me, said, "Where is she?"

"She's on the floor."

I saw a flicker of amusement on Mark's face as he said it.

No response from the front.

I turned beet red, keenly aware that in failing to follow Mark's instructions I had just done something really stupid. Were they straining to keep from laughing in the stilted silence?

As we pulled into traffic, I looked back to see my professionalism lying in a heap on the curb. There was no going back. I knew I had cast my lot with these men. In all my life I had never done what I wanted to do; I had always watched from the sidelines. For once, I needed to follow my heart, take the risk, and see what transpired. I could always return to my controlled life if this didn't work out. My mind harshly criticized my ridiculous behavior, but my soul was utterly and completely satisfied.

I just wished the captain had not been there to see it. Trying to take my mind off the situation, I watched as the men, already dripping with sweat, buttoned their heavy coats up to their chins while the engine forced its way through the streets of lower Manhattan. The temperature inside the cab – exacerbated by the heat of the engine – was easily over a hundred degrees. No civilian could know the grueling demands of their work; my respect for them soared.

Arriving at the call, we all exited the engine since it was too hot to wait inside. Elliot – the driver or "chauffeur" as they called it – stayed with the rig. His job would be to hook the hose up to a hydrant and then feed it to a standpipe system to pump water into the building if need be.

Elliot and I took refuge in the shade of the building. He addressed me. "I knew the mail was there, that's why I told you to sit in the front." I was still feeling pretty embarrassed, and here he was shining a spotlight. But I knew his purpose was to not let me suffer in silence, and he was trying to soften the humiliation by talking to me about it.

I couldn't figure these men out. Why would he bother? Why did he care how I felt? I didn't even know what to say, I was so used to going through life completely invisible.

Another false alarm, and 40 minutes later we left the scene. Mark rearranged the mail so I could have a seat, and we drove back to Engine 32. One of the younger firefighters took sandwich orders. I offered to buy.

"Absolutely not. You're working."

I still expected someone to make fun of me for sitting on the floor, but it was never brought up again. The silence was confusing, as I found myself anticipating the cynical barbs that would never come.

Something had shifted since my last visit. Everyone seemed much more comfortable with my presence. Perhaps it was my travels from Boston, making an effort to do what I had promised, or the effectiveness of the Alexander work – whatever it was – things were different. Steve was first in line for a session this time. He was still as cold as ever, but the fact that he had said yes indicated that the Alexander Technique was, in fact, working. Richie followed right after, with no nudging necessary. Even Ray let down his guard to acknowledge with a simple nod my return to the firehouse, though he continued to decline a session in his heavy Brooklyn accent: "If it ain't broke, don't fix it."

But what touched and surprised me the most was walking into the kitchen and seeing my personal letter to Harry taped to the front of the refrigerator. Stained and partially ripped, it had been up for over two months. I was torn between the awkwardness of having my private thoughts plastered on the refrigerator for all to read, and a childlike joy that somebody thought enough of me to do it.

*　　　*　　　*

There was a new firefighter working the day shift. His name was Freddy Murphy. Fresh out of the academy, he had only six weeks on the job. He seemed to be having some problems with his neck, and I convinced him to try a session. Continually interrupted by alarms, we agreed that I would give him a full session Tuesday night, the next time he would be on duty.

*　　　*　　　*

Monday evening Engine 32 was due to attend a drill on Governor's

Island. (I didn't know what a drill was. I imagined them all in a marching band together.) Elliot drove the crew down to the dock to catch the ferry and returned to Engine 32 alone, allowing for a solid hour and a half of uninterrupted Alexander work. We easily picked up where we left off even though we had not seen each other since February. He was warm and kind, wanting to know what I'd been doing since my last visit. I explained how my experience at the firehouse was changing my life. Something here was so different, and the firefighters were so unlike other men I had known. He listened, but I'm not sure if he understood because he was actually living it. I was on the outside with the objective viewpoint of two distinct worlds: mine and his.

I expressed my frustration regarding Dan and his insistence that it was never a good time to come to New York … that he had continually put me off with how "hectic" it was there. Elliot, too, had the same response as Harry, "It's always hectic."

He was quiet for a moment, and said, "You like Dan, don't you."

I looked at him like he was nuts. I explained in no uncertain terms that I was not looking to get involved with anybody here. I wanted to be a firefighter myself. I wanted the right to be able to work in this firehouse and live in this kind of energy. That's what I wanted. And it was killing me that I would never be able to.

Elliot simply said, "Yes, but who holds you?"

I stuffed down the emotions his question raised. "Oh, Elliot, let's not go there."

He told me a story about a guy named Eric who showed up at Ground Zero the day after 9/11. With his permission, I have reprinted the website version – *Elliot's Story* – here:

*   *   *

After the first day's search, a frustrating and agonizing day of hoping and praying, digging and searching, thinking, sweating, and tearing, it was now time to do it all again. We loaded up supplies: Flashlights, knee pads, tools, and even more determination.

There were about eight of us. We made our way to the back of the World Financial building, which had become our access to the mountainous unknown of One World Trade, where our four brothers were located. We were there without an officer, and so had to be cautious lest we get busted. If caught, we would have been denied access and told to leave. But we weren't about to be denied this mission. So we climbed up four or five stories of indistinguishable mangled debris and massive I-beams that now covered what was West Street.

We still had to make our way a half mile across, down, over and up regions that had barely been explored. We found a crevasse to investigate. Three hours seemed like 20 minutes. We tried to move quietly, hoping to hear some sign of life; just the slightest sound would be all that we needed.

Down in my space, I heard someone call my name. "Elliot."

"Yeah. I'm down here."

It was one of the guys just checking on my whereabouts, as we often did. This territory we were in was deeply unstable. We had to keep tabs on each other to make sure one of us didn't disappear. Then I heard someone yell out my name again, but I couldn't recognize the voice. When I hear any of my guys call me, even in a crowd, I know who it is. We live together. I can even tell who's walking across the second or third floor of the firehouse just by their pace or weight or bounce. Especially Sean's step. Sean Michaels had a very distinctive gait. He would make two sounds when he walked because the front and back of his shoes would fall in one step. But I could not make out this voice.

It called me again, saying, "Elliot, they're calling you."

"I'm OK. I'm right down here," I answered.

When I got a chance, I looked up to see who had called. It was one of the many people wearing hard hats and searching for life, as we were. We decided to move to a nearby area, and this stocky rescuer of African decent, about my size and complexion, headed that way also. Once in speaking distance I asked him, "Who are you with?"

"Carpenters Union, Local 608."

As we moved, I no longer saw anyone except men from our group and this man. "Listen," I said. "Why don't you stay with us?"

"Okay," he said, and we stuck together as we searched. Later on, the Chiefs caught up to us. They told us we had to leave the area; they were shutting down the operation for the night. What little light that had been available was now extinguished, and it was then that I realized how dangerous this uncharted territory was.

We decided to return to this spot in the morning, since we believed it was near the area Matt Dunn, a survivor, described as the last known location of the missing men. As we were leaving, it dawned on me that I didn't see any other carpenters. So I asked him again,

"Who are you with?"

"The carpenters," he answered with that tone like, "The carpenters, stupid!"

But I explained I had meant who was here with him.

He said "No one."

"You're here by yourself?"

"Yes."

"Why are you, like, here?" I inquired.

"To do whatever I can do to help."

"Do you have any family or anyone that might be caught in this?"

"No."

"So you're just here to help look?"

"Yes," he replied matter-of-factly.

I'm thinking to myself: Wow, this guy's just here alone to give all of himself to this horrendous situation. I was thoroughly impressed. Being out there alone is definitely not the move. Something happens to you out there without a buddy and you're gone. No one's going to miss you.

"Where are you headed now?" I asked.

"Home, I guess."

"Where do you live?"

"The Bronx."

"The Bronx?" I asked. "How long have you been down here?"

"Since this morning," he said. "And I'll be back in the morning."

"Yo, look. My name's Elliot. You're welcome to come back to the firehouse with us for the night. We've got a bunk for you. You can wash up and whatever else you need. What you're doing is phenomenal. What's your name?"

"Eric."

Eric hung with the men of Engine 32 for the next six days, without returning to his lovely wife and young son. Finally, after he did go home, he could no longer return to the site because unofficial rescuers were no longer cleared to be there. We hadn't known this right away because, for the first seven days of searching, our firehouse was without electricity or telephones. We didn't even become aware of the phrase 'Ground Zero' until more than a week after the event.

But Eric was there, in the thick of it, from the very first day of the disaster. Down in these extremely dangerous voids like a mole, on his own. He felt the need to act ... to give whatever he could for the welfare of unknown others. He is not a firefighter, but a rescuer. And we firefighters of Thirty-Two Engine recognize Eric as a hero and a brother.

*   *   *

Elliot and I were interrupted by the arrival of another engine company, "relocated" here to cover the neighborhood while Engine 32 was away. Elliot introduced me as a massage therapist and, of course, I experienced the same chilly reception initially received from E-32. They all ignored me. Finally one man relented, explaining that he couldn't turn his neck without significant pain, and asked if I could do anything for him. These were the moments I lived for.

"Do you want me to fix it?"

He was thrown by my confidence.

"Well, sure, I guess."

With the others watching, I put my hands on his back and chest. In two minutes he could freely move his neck without pain. He was incredulous and, best of all, I could feel his attitude towards me change to one of respect. I loved the fact that my physical appearance no longer meant anything here. Only my actions mattered, and through them I was rewarded with acceptance. Elliot sat back and smiled, enjoying the show.

The call came in to pick up the guys from the ferry, so Elliot drove me up the street to where the taxis were plentiful. He asked me to call when I arrived safely at the condo. I protested, saying it was too late to call the firehouse; I didn't want to disturb them. He fixed me with an authoritative look and repeated the order. I couldn't remember a time when anyone cared whether I got home or not, and it felt good.

Not one of the men had mentioned my perfume, and I made a mental note to double the dose tomorrow.

TUESDAY, MAY 7, 2002

Building inspections were scheduled from 10 a.m. to 1 p.m., so I showed up at the firehouse around 2. Dan arrived before his shift started so I would be able to work with him without interruption. Since Elliot made that comment, I was slightly uncomfortable around him. If I was interested in Dan, I didn't know it and didn't want to know it. These were clients and I had a job to do. I was ethically bound and wasn't about to screw it all up by bringing "love" into it. At this point, I could resist a man, but not a ride in a fire engine.

While I was working with Dan, a handsome young firefighter with beautiful eyes and amazing energy dashed into the room. He skidded to a stop like a cartoon character, and said excitedly, "Oooh, massage! Can I watch?"

Dan sat up quickly and barked, "No, you can't watch!" The kid disappeared as fast as he entered.

Such a funny moment seemed incongruous set against the black frame of 9/11, and yet I was starting to understand that these men were intentionally refusing to go into the pain. It was there, but no one was wallowing in it. If they were going to survive and get through this, they needed to stay focused in the moment. They would not let each other down. It was a demonstration of commitment to each other, the likes of which I suspected I would never again witness in my lifetime. I was watching history being made, and I knew it. I was not sure why I had been given the privilege of being here, but I was grateful and humbled. The important thing now was to prove my abilities and the value of my work so that Dan would never question my being here again. I was very pleased when, after we were done, I overheard him tell the other guys that he felt so good he was going to be really nice to them for the rest of the evening.

*   *   *

We were sitting around the dining table. It was a little after six, time to go out for groceries. Steve and Richie were there, and Manny and Matt Dunn had come on for the night tour. I realized that Freddy Murphy, the young kid with the neck pain was not there. I didn't normally speak when I was in the presence of all the men, so they were surprised when I asked, "Isn't Freddy supposed to be working tonight?"

Somebody said he'd been detailed (substituting) over at Engine 7.

"Oh, no! I need to work on him. I promised that I would."

You could see the wheels turning in Richie's brain, as he offered a solution. "We can take you over to Engine 7. You can work on him there."

I was not comfortable with the idea, as it meant meeting with new (and distant) firemen in a new firehouse. It had been hard enough getting accepted at this one. "You can do this?"

Richie said, "Here's what you do. We'll drop you off a half block from the firehouse, you knock on the door and tell them you are Freddy's personal massage therapist and you're there to give him a session."

Steve joined in. "And if anybody else asks you for a massage, you say no, that you only work with the men from Engine 32."

"And that when Engine 32 gets detailed anywhere in the city, we send you out to take care of them," Richie added.

I hesitated, knowing this was another step away from professional. Yet, I realized they wanted to play a joke on Freddy, and without me it wasn't going to happen. A good laugh was good therapy, so I relented.

Matt turned to Dan. "Alright with you, Cap?"

Dan shrugged an okay.

I asked, "How will I get back?"

Richie suggested, "We'll go do the grocery run, and then come back to get you. Will that give you enough time to work on him?"

I said it would, even though it wouldn't. I was a little scared to be outside the boundaries of Engine 32. Old fears of abandonment rose to the surface. What if they didn't come back for me? Or even worse, what if they were trying to get rid of me by passing me off to another firehouse?

Off we went. We drove up to Duane Street and as we approached the firehouse, Richie relayed that their rig was not in. Boy, was I relieved. We continued on to the grocery store. As we pulled into the parking lot, we saw Engine 7 parked near the door just as the firefighters returned to their rig with the groceries. Matt pulled up right next to them. Richie and Steve got out, but the noise of the two diesel engines obscured the conversation.

Next thing I knew, my name was being called, my door was being opened, and Richie was escorting me around Engine 32 and up into Engine 7.

Freddy, sitting across the console, offered a sheepish laugh. "Boy, you sure are persistent."

After an awkwardly quiet ride back to Duane Street, we backed into the enormous apparatus floor of Engine 7 (it could easily hold three or four engines), and Freddy dutifully took me upstairs to a room where we could work. I had no idea of the position he was in now. No one told me that Freddy was a probie – a first year firefighter – and there were specific traditions that ruled what they could and could not do. The session I was about to provide was absolutely forbidden for a probie.

It wasn't long before a very tall firefighter named Chris entered the room and stood there, hands on hips, watching us.

"There's something fuckin' wrong with this picture," he said, and Freddy knew what he meant, though I didn't. He continued. "An experienced firefighter is in the room and the probie's getting a massage? There's something fuckin' wrong with this picture." He left.

"They're going to kill me," Freddy said.

An announcement came over the loudspeaker. "Murphy, department phone. Murphy, department phone." We both groaned, and he got up and left to take the call. Was I ever going to complete a session with this kid?

Chris came running into the room, and jumped on the bed. "Pretend you're working on me. When he picks up that phone, all he's gonna hear is a dial tone."

Freddy returned and found Chris in his place, and we all had a good laugh. Chris was sport enough to let me continue to work on Freddy.

One of the officers came in to watch, and asked if I was going around to all the firehouses.

Remembering Richie's instructions, I replied that I was working only with the men of Engine 32, and when they were detailed anywhere in the city I was sent to work on them. The officer was visibly stunned; I wished the guys from E-32 had been there to see the look on his face. Believing I was serious, he said, "You mean I have to get detailed to Engine 32 in order for you to work on me?"

But there was something about the way he said it that made me quit the joke. I asked what was going on with him. He had injured his shoulder in an explosion at a fire in November. I apologized for the foolishness and assured him I would work with him after I finished with Freddy.

The lieutenant came in after Freddy left the room. His shoulder was really screwed-up, and I knew I would need a full hour with him. But I wasn't going to get it. Ten minutes later: BING BONG! They all took off, and there I was ... alone at Engine 7.

I marveled to find myself in the very same firehouse that was featured in the documentary film *9/11*. Why was I being given such a personal window onto this history?

I returned downstairs to the apparatus floor. Engine 32 was due back from the grocery run at any moment, and they would never find me upstairs. I sat patiently and waited.

And waited.

They didn't come.

I began to doubt the wisdom of what I had done. It was a bit déclassé, jumping from one fire engine into another. Although I was certain of the value of my work, I couldn't be sure that these guys were taking it (or me) seriously.

Forty-five minutes passed. It was now 9:00. I withered inside and thought that maybe I should take matters into my own hands and just leave. I didn't want Engine 7 returning only to find that Engine 32 hadn't bothered to come back for me. But there was something about the act of taking care of myself, and never being able to depend on anyone, that didn't apply here. It would mean I didn't trust them.

This was the moment. Either I was going to trust them to do what they said they would do, or not. If I was crazy enough to sit on the floor of a fire engine and suffer that embarrassment, I might as well see this all the way through to what I prayed would not be a humiliating ending.

I heard, then saw, a fire engine pull up to the door, but it was only Engine 7 returning. They backed into the garage, and the lieutenant got out of the rig and walked toward me. He explained that Engine 32 had responded to the same fire, and they were still there. Their plan was to return to their firehouse, cook the meal and come get me when dinner was ready. This would allow me more time to work at Engine 7.

Totally unprepared for this, an old knot somewhere in my stomach released. They were not going to let me down. Is it possible they understood how important this was? No words had been spoken by the men of E-32 regarding my purpose here, but it was clearly being acknowledged. The lieutenant and I went back upstairs to continue his bodywork. He said that his shoulder already felt better.

Finally, the call came over the P.A. "Jessica, your ride is here."

It was unseemly to hear my name blasting so officially throughout the building. I promised the lieutenant I would stop by on Wednesday to work with him again after I finished with Engine 32.

We went downstairs. Engine 32 had pulled up on the apron. The men were gathered outside, talking with the firemen of Engine 7. It seemed routine. In spite of a fire and a dinner that was now hours late, they had come back for me. We didn't speak as I approached; it wasn't necessary. *They had come back for me.* It was so simple, but profoundly satisfying in a way that brought tears to my eyes. I turned my attention to the star-lit sky, both to give thanks and to keep them from seeing how much it mattered to me.

A crowning moment emerged from a lull in their conversation, when Steve turned to me and asked if I played cards. It was an invitation that indicated he had forgiven my football inadequacies.

"Poker?" I asked.

They all laughed. I don't know why.

WEDNESDAY, MAY 8, 2002

Andy nearly had a heart attack when I told him I had worked with Freddy over at Engine 7.

"You gave a probie a massage?!!!"

I was stunned by the strength of his reaction. He did not see the humor at all. He launched into a serious lecture about a firefighter's probationary period, and the purpose behind it: To build character, to learn to be a part of a team.

"If I'm going to trust this guy with my life, I want to know who he is, how he's going to react at a fire if something goes wrong. He doesn't get to walk in here and be accepted. He has to show us his commitment to the job and to the house, and prove it before we let him in. He has to earn our respect. We don't just give it to him."

Finally, I understood my strong attraction to this place:

First, what was going on in this firehouse was not an accident. There was a structure here that allowed young men without any life experience to gain self-esteem through earning acceptance into the firehouse.

Second, by keeping my mouth shut, my ears open, and working hard, I had accidentally stumbled into that structure;

Third, it was starting to work for me. My own sense of worth was being restored from scratch. The harder I worked and demonstrated my commitment, the more respect I earned. Once it was earned, they continued to honor it. I didn't have to start over again each time I arrived. And it was this newfound self-esteem that was making me less afraid of being in the world … less afraid of trying something new.

Lastly, unlike my mother, I knew Andy liked me even though I had erred. The tone in his voice throughout his reprimand was that of a loving father; it did not serve to drive me into further self-loathing. It was filled with education and forgiveness.

I felt all of five-years-old, looking up to all these men to help me find my way. This firehouse was more nurturing than any household I had ever been in. I was not crazy to want to be here. This was the most intelligent thing I had ever done for myself. I was being re-parented from the ground up. Never having experienced self-esteem, I wasn't really sure what it felt like. But now I knew. I can only describe it thusly: It is real, like a mountain fortress growing slowly inside you, lifting you up higher and higher from within. There is no fear. It prompts you to give more and do more, because you are safe inside. It is absolutely wonderful.

I apologized to Andy, acknowledging my mistake and promising not to do it again. Then it was time for me to take charge.

Andy was suffering from severe respiratory problems brought on by the dust at Ground Zero. Almost every morning, he would cough until he vomited. He was told he had breathed fiberglass into his lungs, but that they were short fibers – not long fibers – thought to be less dangerous.

"And that is better … how?" I asked.

"It's all bullshit," he said. "They're just telling us bullshit." The FDNY Medical Department had determined if the affected men hadn't lost more than 20% of their lung capacity, they were still considered fit for duty.

I didn't know what to say. How can you lose 20% of your lung capacity and feel normal?

While I couldn't reverse the scarring in his lungs, I knew I could release the muscles in the rib cage to allow for more lung expansion. This produced a very calming effect as the body sensed it was getting more air. It wasn't much, but it was something.

Lunch was announced over the intercom, but when I looked to Andy, he indicated it was fine for us to continue. For any of them to miss a meal, I must be doing some good. I worried if they didn't eat when they had the time; an alarm might come in, and they could go hungry for hours. So I said, "Ten more minutes." He agreed.

Afterwards we went down to lunch, quietly taking our places at the table under the conversation in progress. There were guests today; someone's cousin and elderly uncle were visiting, both retired firefighters. Introductions were made. Somebody referred to Lou as Mike. Confused, I turned to Ray and whispered underneath the dialogue going on. "Isn't his name Lou?"

"It's Mike, but we call him 'Lou' … short for 'lieutenant'."

It finally dawned on me why an inordinate number of guys in this firehouse were named Lou.

After lunch, the six younger men took the rig out to tour Ground Zero. I was left with the elderly fireman. His name was Ed. He told me that he thought I was one of the firefighters … that I seemed to fit in. I loved hearing that, especially since I had finally discovered and understood the structure.

I asked him if he had any funny stories to tell about being a fireman and, without hesitation, he started right in. The funniest tale was about a fire at a funeral home. The place was filled with smoke, and he was feeling his way through the room when he came across a casket. There was a body in it, and he thought it would be a terrible thing for the family to have the body burn, so he tries to pull the body out of the casket, but the corpse is too heavy. He's got it half way out of the coffin, and is seriously struggling with it. So he starts calling. "Guys, c'mere, I need some help," and the firemen find their way over to him and say, "Eddy, what are you doing? You can't save him, he's already dead!" "Put him back!" He said they never let him hear the end of that one, "You should put in for a medal for that, Ed, saving a dead guy." I was on the floor laughing.

After one more session with Matt, I was done with Engine 32. John O'Brien, the senior man who did not like me in the firehouse was working the next day so, in keeping with my intention to respect his wishes, I did not extend my stay.

I stopped by Engine 7 to check on the lieutenant as promised, but he had been detailed to another firehouse at the last minute. Chris was there, and he declined my offer of a session, choosing instead to razz me by saying he wouldn't dream of taking a massage from "Engine 32's *exclusive* massage therapist."

BOSTON, MASSACHUSETTS
MAY 10, 2002

A few days later after arriving home, I was surprised to receive the following email:

----

on 5/10/02 2:02 PM, Dan Fullerton at danfullerton@yahoo.com wrote:

Hi,

Hope your trip back was OK.

What did you think of Engine 7? That firehouse is a little different than Engine 32. There are a lot of different guys there since I worked there. It's still a good house though.

The work you did on me seems to have helped; I'm sleeping better at least.

If you get a chance drop me a note so I know you got back OK.

Talk to you soon. Take care.
DF

After the embarrassing moment on the floor of the fire engine, I was not sure where I stood with him. The letter eased my anxiety completely. Things were okay, he didn't think I was a total loser. Man, these guys were forgiving.

From: Jessica Locke <jessical@shore.net
Date: Fri, 10 May 2002 23:47:05 -0400
To: Dan Fullerton <danfullerton@yahoo.com
Subject: back home

Hi Dan!

I'm so glad to hear you are sleeping better.

Okay, I've been meaning to talk to you about the men. I ask that you keep this letter strictly between us. I write out of concern and committment to them.

The individual sessions have been profoundly moving. Poignant, sad, funny, heavy, you name it. The hectic schedule of the firehouse is such a strange background in which to be doing this work. Always, on the train home, it hits me all at once, the personalities and intelligence and goodness and humor, and I am just brought to "my God, these incredible men." And I feel so honored to have had the opportunity to spend time with you.

As I had hoped, showing up again made for a

major increase in their trust of me. We got a lot more done this time. They, the caretakers, actually let me take care of them. Briefly. A toe in the water here and there. It was a big step for those concerned, and me as well. At what point do these men ever get a chance to let down their duties as firefighters? For each other, for their families, for the public—everyone keeps leaning on them because that's what they are there for. They save us. I think the stress of this is going to take its toll over the next six to eight months. I don't know if I am making a difference in the outcome, but I have to try. I cannot stay up here in Boston and watch this play out and do nothing.

When I first came down in January, I worked with 3 men in 4 hours. So easy and efficient. I thought it was always going to be like that. I was in total shock for the February visit when I found myself at the firehouse til midnight finishing guy number 4. After being there 11 hours. Jeesh! This last visit I deluded myself into thinking I could run a schedule from 2 to 8, work with maybe 3-4 guys, and leave for a relaxing evening in the condo-- ha! But I did do better this trip.

I worry that my presence there for so long is too much, but I don't know how to get to everyone without adapting to the firehouse schedule. I need to know your feelings about it. And I am asking you, how can we make this work? Is it working now the way it is?

Thank you so much for allowing me to come down, Dan, I hope this letter clarifies a few things.

Take care,

Jessica

----

on 5/13/02 2:03 PM, Dan Fullerton at danfullerton@yahoo.com wrote:

Hello again,

I hope your trip back was pleasant.

Your concern for us is so heartwarming. It seems like you are getting an idea of the culture of the Fire Dept. I'll try to explain. A fire company is a very close-knit group of very dedicated and motivated professionals. We think we can do anything. As you have seen we don't just work together we live together. It's like a second family. The teasing and practical jokes are a tradition in firehouses. It serves a purpose. It breaks the monotony of waiting for an alarm or a fire. It's also a way we remind ourselves not to take ourselves too seriously. It keeps our egos in check. The only way we can effectively operate at fires and emergencies is through cooperation; there are no "stars". Every guy has an assignment and depends on every other guy to do his. This concept follows through in everything we do. When we clean the firehouse everybody helps. When we

make the meal everybody lends a hand. You
probably noticed when we go shopping everybody
goes in the store. If one of the guys needs help doing
some work on his house guys will just show up. If I
was stuck somewhere in the middle of the night
because my car broke down all I would have to do
is call the firehouse and I know help would be on the way.

It is because of this that losing members of a company is
so painful. Engine 32 lost FOUR in ONE DAY. At any
other time the rest of the Fire Dept. could rally around
a company like that and lend support. (It's traditional that
if a company loses a member in a line of duty death the
last company to lose someone will help them with funeral
arrangements etc. because they have an understanding of
what they are going through.) But now it's different. So
many companies lost so many men we have to help
ourselves. The counseling unit was geared to help one or
two firehouses if they lost members. Now there are
dozens of companies who lost hundreds. The counselors
are overwhelmed. The Fire Dept. is trying to help but
no one could have predicted this so there was no
contingency plan. They have to figure it out as they go.

It is hard. We need help. We don't know how to ask
for help. People call US for help. That's where people
like you come in. So many people from so many places
have offered so much. We are very grateful but a lot of
the time we don't really know what we need. We have
to figure it out as we go. When I was a new firefighter
a senior guy described operating at a fire as "organized
chaos". That's how things seem now.

Besides the members who were lost on 9/11 there are many still on medical leave, and many who have retired. They can barely hire enough new guys to replace them. We are practically rebuilding the whole Dept. This also makes things difficult. A new guy can't just walk in the door and be accepted. He has to prove himself. He must demonstrate his level of commitment. This of course takes time. This has something to do with what happened at Engine 7 with Freddy. He only has about 6 weeks in the firehouse. That's where the "something's wrong with this picture" comment comes from.

You are right, these are incredible men. I've seen them put aside their pain and grief to do their jobs. But this is the job we've chosen and the job we love and wouldn't trade for anything. As you can tell they haven't lost their sense of humor. They do need to open up a little sometimes. It's not our nature to do this easily. (It's a guy thing) You seem to have the sensitivity to get some of them to do this. Trust comes with time.

I don't think I ever explained how I came to be the Captain in Engine 32. It wasn't by chance. The former captain of Engine 32, Roger Sakowich, is a friend of mine. When Roger was promoted he was very concerned for the guys in Engine 32. They had been through so much. Roger contacted me and asked if I was interested in taking the spot at Engine 32. I knew this would be a difficult situation: taking command of a company that had lost members. But when Roger said to me, "you know how I feel about these guys..." I didn't hesitate. These guys are terrific and we'll get through this together.

I hope you know you are welcome always at Engine 32.
Your visits give us a boost. It's good to know there are
people who care and are willing to do something. Don't
worry about getting to everyone when you are here.
Come when you can, do what you can when you are here.
This is something I had to learn in the FD. The first fire I
went to where we had a fatality I felt as if we failed. It
takes a while to accept the fact you can't save everybody
all the time. I mentioned this to a Chief once and he said,
"We didn't start the fire, we put it out." Simply put, we
do good, we do our best. It's working. You help each
time you come, that's plenty.

I don't know when you were planning to come down again.
Things will be pretty hectic till the end of the month.
We have to clean the firehouse for annual inspection.
We just received a new apparatus and have to get
familiar with the new piece of equipment. Also one
of the members who was lost was identified recently
and his funeral will be at the end of the month. So
as usual give us a call before you make any plans.

Take Care
DF

From: Jessica Locke <jessical@shore.net>
Date: Mon, 20 May 2002 02:37:36 -0400
To: Dan Fullerton <danfullerton@yahoo.com
Subject: May 30th

Hi, Dan,

Thanks for the advice about trying not to save
everybody. I needed to hear it, more than you know.

I heard the news about the closing ceremony on the
30th at the WTC. It doesn't seem from what was said
that it was a closed situation, i.e., that maybe I could get to
Church Street early enough to get a decent view. Please,
if you should hear otherwise, let me know. I'm planning
to come down on the 29th.

The whole thing has put me into a state of I don't know
what. For myself, it's an awful day. A day that says,
"this is over," but it's not; it's the beginning of something
very very hard. I'm not sure I'm ready for it. But whatever
it's going to bring me to, I need to face it.

With great affection for all of you,

Jessica

----

on 5/22/02 2:25 PM, Dan Fullerton at danfullerton@yahoo.com wrote:

Hi,

The wake and funeral for Sean Michaels will be on the 30th
and 31st. It's probably best if you wait till after this time to
come down. I know you are anxious to get down here
and do what you can but I think the guys may need a
little time. I've never been in this situation before

and don't really know what to expect. I'm just going
to try to take this one day at a time and do the best
I can. I promise I'll keep in touch and let you know
whats going on with the guys you know.

Stay well,
DF

----

on 5/27/02 2:24 PM, Dan Fullerton at danfullerton@yahoo.com wrote:

Dear Jessica,

I have a little info about the World Trade Center
closing ceremony. On May 30 a closing ceremony will
be conducted at the site starting at 10:29 AM. I'm
sure the site itself will be extremely crowded with
family members of victims. In each firehouse however,
the apparatus will be pulled out toward the street and
a moment of silence will be observed. An ecumenical
prayer prepared by FDNY Chaplains will be read by one
of the members. The Dept. has encouraged families of
fallen members who are not participating in the
on-site ceremony to participate at the firehouse. I
would expect quite a few people from the neighborhood
will be there. I think you said you wanted to come
down for the closing. If you are, going to the
firehouse might work for you. It's probably better
than standing 6 blocks away from the Trade Center and
not seeing or hearing anything. There may not be
anybody from Engine 32 there though. The wake for Sean
Michaels (who was lost from Eng 32) is that day and the

Funeral the next. There will be guys from other
nearby firehouses manning the company that day. I am
going to try to be there though and then go to the
wake from there.
Hope this helps. Talk to you soon.
DF

From: Jessica Locke <jessical@shore.net
Date: Tue, 28 May 2002 19:43:44 -0400
To: Dan Fullerton <danfullerton@yahoo.com
Subject: Re: WTC ceremony

Dear Dan,

Thank you for telling me about the firehouse service –
that is so beautiful and moving that they are going
to do that all over the city. I've been wondering whether
I could get to the WTC site early enough to see
anything, but I knew I wanted to be in NYC for this
no matter what. I'll come to the firehouse.

It will be nice to see you, but if not, know my thoughts
are with you.

Here we go. I have a feeling this is going to be bigger
than we know.

Jessica

MONCLOVA, OHIO
DECEMBER 18, 1988

My mother was dying. After a mild seizure, a CAT scan revealed a fast-growing tumor in her brain. The doctor announced – rather arrogantly – that she would be dead in three weeks.

In many ways Mom ran our lives with a total disregard for society's rules, so my siblings and I took over the intensive care unit at the hospital without any regard for their regulations. Mom's needs, as always, were law. I settled into the night shift, challenged to sleep in a chair by her bed. When I laid my head on the bed next to hers, the nurse yelled at me. So I slept on the floor. The nurse said I shouldn't do that either, that I had no idea what had been on that floor. It seemed pretty clean to me, having slept with fleas and chickens before.

One day I came into the room to find an aide attaching electrodes all over Mom's scalp and body. "What the hell are you doing?"

The doctors wanted to run more "tests" to see if the cancer had spread.

"What does it matter? You said she would be dead in three weeks, so why are you doing this?"

I didn't wait for the answer. "Stop it." I called my brothers and – like the pivotal scene in the movie High Noon – we walked down the long hallway together to inform the doctor that we were taking her out of there. We were taking her home.

To instill fear, the doctor reacted, "You don't know what you're doing.

What are you going to do if she has a seizure? Have you ever seen anyone have a seizure?" My brother Ben looked him right in the eye and said, "We'll handle it." *I couldn't have been prouder of him. We were used to doing things we had never done before.*

And so we brought her home and set up a hospital bed in the living room. Again, I took the night shift, sleeping on the couch next to her, and turning her every two hours to avoid bedsores. Dad turned the thermostat up to seventy two degrees. I had no idea the house could actually get that warm.

Mom said, "Now I know I'm dying. Your father turned up the heat."

One night as I was drifting off, I looked up to see the silhouette of a man lurking in the doorway to the hall. I jerked awake in absolute terror, uncertain if I had seen it or dreamed it.

The next night I saw him again. Even in my waking hours, I was haunted by the image. I didn't tell anyone; I rationalized that I was delusional from stress and sleep deprivation. Yet, I sensed that with Mom's impending departure he was coming to get me, waiting in the hall until she was gone. Again, on the third night, he returned. Now I was sure he was real. I grew more and more alarmed, until the terror finally took hold of me. Although I wanted to be with my mother, I returned to Boston five days before she died.

Just as she was leaving, I came to realize that I loved her, that I had always loved her. Through her illness and death I discovered anew, deep within me, the God-given connection between a child and its mother. The instinctual response designed by nature to protect me – which she managed to obstruct at every turn – now returned to leave me helpless and forever unable to let go of her and grow on my own. She had created me to serve her needs. Now, my Creator was dead. I didn't know what to do.

MANHATTAN
WEDNESDAY, MAY 29, 2002

I arrived at Penn Station late in the afternoon, met my friend Leon for dinner, and then went back to the condo for a quiet and reflective evening alone. Tomorrow's weather report was not good: The observance at Ground Zero would be held in temperatures well into the 90's. It was going to be hellish in the gritty, sauna-like subway, so I decided to take an air-conditioned cab down to Engine 32. That would help save the nice dress I would be wearing for the ceremony, which was to start promptly at 10:29 a.m., the time the second of the two towers fell.

The day dawned hot and steamy as predicted, and I reluctantly left the cool lobby of the hotel at 9:15. I hailed a cab on Seventh Avenue, but after just half a block realized its air-conditioning was not working. Rather than risk being late, I decided it was best to just deal with it. My hair drooped in the heat and humidity.

The traffic was terrible. The huge memorial service was wreaking havoc for anyone headed to lower Manhattan by car. We crawled along; I could have walked faster. It was now ten o'clock, and we had only arrived at Union Square. I grew nervous. The driver recommended I take the subway, so he dropped me off in front of the unfamiliar East Side line. My map was at home, and now I needed it badly. Uneasiness crept into my soul; nothing was going right.

I asked a few subway riders where I should get off for the Trade

Center, but they didn't speak English. Someone nearby overheard, said they were going to the ceremony, and offered to help me make the correct transfer. I ran up the stairs, temporarily disoriented. It took me awhile to determine which way to go. I dashed down the street as quickly as I could.

There was a young police officer at the corner. Of course, anyone running was a reason for his interest to be piqued, although I was fairly certain that the pearls around my neck would not register me as a terrorist threat. As I blazed past, so as not to alarm him, I called out, "I'm late for the ceremony at the firehouse!" I turned the corner and stopped short.

The place was deserted. No fire engine, no people standing around, no ceremony, no nothing. The air suddenly grew heavy; it was 10:29. I approached the firehouse, and looked into the window. The engine was gone.

Thus far, the most important lesson learned from Engine 32 was to always trust the flow of things, to surrender to what was. This, however, was excruciating. Tears welled up and I fought them back, not sure what to do or where to go.

At that moment the door of the building which stood adjacent to the firehouse, opened. About a dozen people exited and started ringing the doorbell to Engine 32. I didn't bother to point out that the engine was gone and there would be nobody there. But to my surprise, the big red door began to rise and there stood John "you left your jewelry upstairs" O'Brien, the senior guy who had been so disapproving of my presence at the firehouse back in February.

Inwardly I shrank. I approached him and reintroduced myself, explaining to him that I was there was because Dan had told me about the ceremony. He explained that there had been a last minute change of plans. They had all gone to Engine 14 for breakfast prior to the service, and I would find them there.

It was out of the question. I would not barge in on them on an important day like this when I hadn't even been invited.

Sweaty, disheveled, demoralized and defeated, I asked John if I could watch the ceremony on the television in the dining area. He said that would be fine. I walked through the garage to the back, reeling from the fact that I had come all this way to be part of this important ceremony, and now I was going to end up watching it on television by myself.

I walked into the dining area and stared at the screen. I sat down at the table, but the scene in front of me was so powerful that out of respect I had to stand again. The people from the building next door were talking loudly outside on the apparatus floor, so I shut the door.

I was alone, watching a funeral for two lost buildings, a ceremony marking the end of hope without redemption. The last beam, dressed in black cloth and covered with an American flag, was escorted from the arena. It was all over. There were no more people to be found; they had disappeared without a trace. Unfinished lives, gone in a meaningless moment. Why?

My eyes wandered to the blackboard next to the television. Scrawled in large letters, it read:

We need a new mop head

Here in this grief-laden firehouse, sorrow would remain subservient to their job. They needed a new mop head.

Then, suddenly, in that unguarded moment, I felt their pain; in the walls, the floor, in each book, pen, paper, memo, picture, chair, table, couch. The amount of pain I sensed was excruciating. It was an unbearable weight, and I understood now why they could not speak of it. How could a firehouse cry?

The silent siren rose upon itself, higher and higher. It began to magnify, and resonate, and reverberate through the walls. It shattered every makeshift defense I had assembled within my being. My body shifted and began to experience odd sensations which started at my feet. A tingling sensation rose up through my legs and into my trunk, up my neck, until it finally reached the long-frozen divide in my head. The two sides of my brain began to alter, to melt. I felt myself sinking down into my body as though I had never been in it before. It was a vile feeling; disgusting, foul, and wrought with sexual overtones. Ugly and evil, it encompassed all the reasons I had long ago pulled away, down, inside and out of my self.

Last night, I had a dream I was dancing with my cat, her front paws in my hands.

My mother enters the room, and with one swift chop of her hand cuts the cat in half. The cat falls to the rug in two pieces. The bottom half of the cat turns black, the upper part turns white. The front half of the cat struggles forward, dragging itself by its claws, as bloody entrails pour out. In horror and grief I look to my mother to do something. Do something. The cat must be taken to the vet and sewn back together, or it will die. But she just stands there, a look of sick satisfaction on her face. I cannot speak, nor can I make any move, lest she do the same to me.

No! I didn't want the cat sewn back together. I didn't want to be rejoined with my evil, ugly self again. Especially not now. Not here. Not in this firehouse. I would never want to bring such horrible energy here; I loved it so much. I had worked hard to keep my broken side buried, to bring only the best of me to this place.

But there was no stopping it. The two warring entities within me continued their clash, and I saw my mother's hand slicing the World Trade Center towers in half, toppling them to the ground. I didn't know what was happening. Though I tried desperately to fight it, I began to sob uncontrollably.

John came in, assessed the situation and said, "Are you okay, little lady?"

I choked out, "Now what?! Now what?!" This horrific state perfectly – and psychically – matched the utter hopelessness of all those people standing at Ground Zero for whom there would be no peace, not even a body to bury for closure. My divided mind settled into the rubble as I was enveloped in total blackness and despair.

Then, an invisible presence restrained me from speaking, firmly informing me that the cover-up was over. There would be no more faking or pretense. My dark side was being summoned to account for herself.

I froze, staring into John's eyes, pleading for him not to see me. Of all the men to be with at a time like this! This man had always known about the evil within me. That was why he hadn't wanted me here; he was smarter than all the others. He saw right through me. I hadn't fooled him for a second.

I looked deeper within to find a way to fake through this, but there was only the crying and a terrible fear of being seen. I had an appalling awareness of how bad I truly was; a murky hole of filth and slime, and now, up rose a terrible shame for a deed I couldn't claim. Surely, John would tell the others, and I would not be allowed to return.

"Can I make you a cup of tea?"

What? I couldn't answer him. His question made no sense, and served only to intensify my tears. Here I was, standing before him, exposed, slimy, and deceptive. I was not here to take care of these men; they didn't need help. I was here because I needed them. Why wasn't he telling me I was a slut because of this need, that I was no good. Why was he being nice to this filthy presence, instead of wiping his feet on me like all the other men who had gotten this close? Tea? Tea?!

He waited a moment longer for any response. No words would come. "I'll go make you a cup of tea."

He went to the kitchen and I heard him fussing about, putting a kettle on the stove, opening and closing the cabinets. Moments later, he brought a cup of tea, a spoon, a napkin, and a sugar dispenser and laid it before me like a kindly grandmother. I was bewildered.

We were like actors in a play. I sat there, waiting for my cue without knowing what my next line would be, just staring down through a blur of tears at the cup, the spoon, the napkin. Why would this man, who didn't particularly want me in his firehouse, bother with a napkin?

Just then, in the seemingly broken recesses of my mind, I began to experience amazing clarity. Though the events of 9/11 and my past were somehow strangely intertwined, that was not why I was weeping. I was, instead, crying for this firehouse. I was crying because within these walls, at every turn, I had been treated with extreme respect, courtesy, empathy, and compassion. These men had endless opportunity to laugh at me, abuse me, crush me, and take sexual advantage of me. Instead, they chose to hold out their hand and lift me up ... to offer me a place to be safe. And in that safety, they gave me a place to grow.

I had waited and waited for them to see the awful side of me, but they would have none of it. The only thing that mattered here was action. But my actions had been the same my entire life, so was it possible then, that the rest of the world had been wrong about me? That *I* had been wrong about me?

I had been searching for this firehouse all my life. It was a chance to start over, to be pure again. It was a place where I could admit that I didn't know how to live in this world, that I needed help. If you are in trouble, ring the bell, any firefighter will help you. It was true. My God, it was true. These men of Engine 32 were the living embodiment of goodness. The grace of God, now, was physically real. For all the evil consuming the world, there were still men who would stand up and do the right thing. They would not give in nor lower their standards no matter how difficult, no matter how great the pain or loss, no matter what.

The course of the battle in my head was now transformed forever in my heart and soul. It was painful to cry out with relief at the possibility of being whole. I could feel pain again. I could actually feel my feelings. Standing in this clarity, it was obvious to me that I was not evil, nor ugly, just hurt ... *very* hurt. By who or what, I didn't know, nor did it matter now. A huge burden was lifting from my heart.

Still tearful, I apologized to John for falling apart. He assured me it was okay, and again suggested that I might find the guys at Engine 14, but I wanted to go to the World Trade Center site. As he escorted me to the door, he said that perhaps some of the men would be there. I hoped not. I didn't want any of them to see me like this, my eyes red and swollen, my makeup gone.

I asked him to tell Dan I stopped by, and he replied, "I'll tell him you bawled all over our dining room table." It was the truth, and it made me laugh.

I started up the street, but then turned back. John was still standing outside. "I just want to say that knowing you guys has been the best thing that ever happened to me. You are all just great." He seemed to swallow hard. Maybe not. I just remember looking straight into his eyes and meaning every word.

I headed back up the street, tears still streaming down my face, but I didn't care. I was whole. I was real. I wasn't hiding anything anymore. The cop, still on duty, took one look at me and said, "Are you okay?" I must have looked like hell, but I assured him I was fine. He said kindly, "Do you need a hug?"

Now I was crying and laughing. Up to this point my entire life had been so incredibly wrong. The mean and cruel streets of New York had simply never existed except in my head. The rapists, robbers and murderers were supposedly everywhere, and I get a guy asking me if I need a hug. It was so perfect. There was no way to explain to him that I was alright. To put in plain words that something wonderful had happened to me in that firehouse would surely have made no sense to him since I could barely understand it myself. I wanted to go to the World Trade Center. I found myself running to get there ... to go home.

As I got closer, I saw hundreds of people streaming out and away: Families, and police and firefighters in their dress blues. Many met my eyes and we nodded. There was a passionate sense of unity, of being part of a community of people who understood the worst kind of pain. This was my country, and these were my people. We all were having the same experience, and I felt safe and loved.

I walked past the memorial fence at St. Paul's and every poster, every candle, every flower emanated its personal contribution to the aura of grief. How many times had I passed this fence, unable to take it in? Now I could feel it. I was in the world, and everything was okay.

Firefighter after firefighter passed me and each one met my eyes and nodded or spoke a greeting. For 50 years I had managed to go through life without anyone noticing, and now I was seen in all my pain. Me. The inside me. Just me.

I stood in silent respect at the site, drinking in the profound sense of connection to everything and everyone. I was dazed. The shock of this new-found experience temporarily silenced any ability to consciously think.

Finally, I left and rode the subway toward home. Time and space were suspended, and I didn't snap to until long past my stop, at 72nd Street. I walked up and out toward Central Park, planning to walk back downtown to the condo.

There was something terribly familiar about the building at the end of the block. I stood there, struggling to remember where I had seen it before. The Dakota. I recognized the doorway where John Lennon was murdered; John of my beloved Beatles. I recall being unable to cry. The hurt of losing him had been separated and buried, too. How strange, I thought, that I should end up here on this day. It was as though all the grief of my life was releasing from stoic depths to wash through me at last, all at once.

At that moment, the front gate opened. I was astonished to see Yoko Ono step out and walk toward me, headed toward a black car waiting at the curb. Our eyes met in a silent acknowledgement of universal pain.

She, too, nodded at me, and then entered the car.

*   *   *

That evening, on the train back to Boston, I curled up across two seats and covered myself with my coat against the draft of the air conditioning. I adjusted the coat over my face, so I could continue to cry freely.

As I thought back on the day, I knew a miracle had happened. The "voice" – which I had perceived as an omen of my death – had guided me to Ground Zero to what had become a new life. I mentally reviewed past visits to Engine 32, and as I recalled each cherished memory, an image began to appear in my mind. There were not 25 individual men, but one man assembled in spirit out of the unity of them all. This one man, a warm glowing entity, also embodied those lost on 9/11. He had been with me from the beginning and was, in fact, the voice that insisted I travel to New York. My new vulnerability allowed me to acknowledge his presence. I realized now that these firefighters, on some level, knew everything about me. They knew I had been broken and badly hurt.

I had not been brought here – summoned down here – to save these men. I had been called down here so they could save me.

My body shook beneath my coat as I forced the final sobs into silence. I was embarrassed and ashamed that I had once believed them incapable of such loving awareness. They had opened their door to me without ever asking for – nor expecting – a thing, and they were going to let me stay for as long as I needed them.

I was enveloped in a timeless moment of being moved closer to God. I vigorously renewed my promise to take care of them; to work harder, stay longer, and come down more often. I would never be able to repay them for saving my life.

145

Don't Look For Me Anymore

Don't look for me anymore
It's late and you're tired
Your feet ache standing atop the ruins of our twins
Day after day searching for a trace of me
Your eyes burning red
Your hands cut bleeding sifting through rock
and your back crooked from endless hours of labor ...

It's my turn
I'm worried about you
Watching as you sift through the ruins of what was
Day after day in the soot and the rain
I ache knowing you suffer my death

Rest in knowing that my blood lies in the cracks and crevices
of these great lands I love so much ...

Don't look for me anymore
Hold my children as I would
Hold my sisters and brothers for me
since I can't bring them up with the same
love you gave me
And I'll rest assured, knowing
You're watching my children
Don't look for me anymore
Go home and rest.

poem by Alicia E. Vasquez, dated September 14, 2001
(left on a fence near the World Trade Center site)

BOSTON, MASSACHUSETTS
JUNE 8, 2002

Seeing an incoming email from Dan prompted me to hold my breath. The subject: *Sorry.* I was afraid to open it; perhaps falling apart at Engine 32 had raised questions about my ability to continue the work.

From: Dan Fullerton <danfullerton@yahoo.com>
Date: Sat, 8 Jun 2002 08:11:43 -0700 (PDT)
To: Jessica Locke <jessical@shore.net>
Subject: sorry

Dear Jessica,

Forgive me for not writing sooner. I'm sorry I missed you on the 30th. I heard you had a difficult time. Don't worry about losing it in the firehouse. It was probably something you needed to do. You are not the first or, most likely, the last.

Thanks for the poem. Sometimes we need to be reminded life is for the living.

Hang in there.

DF

With that compassionate and understanding letter, I finally gave in and admitted I had fallen in love with Engine 32. I was in love with a firehouse. Never before had I entered into any relationship without having an exit strategy, but now I found myself moving toward this, praying there would be no way out. There was a beauty, nobility and humility to the truth of it that repeatedly brought tears to my eyes and irresistible warmth to my heart. It was more real and more satisfying than any relationship I'd had before. Its basis was safety, trust, and reliability. I felt it grow richer with each visit, deepening our connection to each other. I could not lose this, for now I could not imagine my life without them.

* * *

The air conditioning broke down in my 17 year old Audi during one of the hottest summers on record. An $800 repair, I certainly couldn't afford to fix it, but then, there would be no point in fixing it since the car was on its last legs. I put ice packs around my neck and drove. Whenever I made a sharp turn, a loud clicking noise emanated from the left front wheel. The gas station attendant told me the sound indicated an impending break in the axle. So, I drove 25 miles per hour, watched zealously for potholes, and never went on the highway. I begged the car for just six more months. Please, just hang on for six more months so I can keep my commitment to Engine 32. It is a testament to German engineering that it did.

* * *

Every once in a while I would clean up the landlord's yard, weed the garden, etc., just to be nice, make a contribution. This year, as I stood there in the hot sun raking, the landlord arrived home and sauntered through the garden, surveyed my work and said, "You took out one of my mint plants."

In the old days, I would have crumbled under the accusation and assumed I must have done it. But I knew I didn't. My ex-husband was an obsessive gardener and, consequently, I learned the Latin name of every common plant in the Northeast. I would have recognized a mint plant in a second. I took in a deep breath and said, "No, I didn't."

He said, "Yes, you did. There were three of those plants, and now there are only two."

I said, again, firmly, louder, "Well, that may be so, but I didn't do it." Never again would I let something be said that wasn't true. Never again. He walked away, and in my mind I said for him, "... and thank you so much for raking the yard, Jessica."

This self-esteem stuff was amazing, once it was in your system. What was the point of working for people who did not appreciate what you did for them? I suppose this was another pattern courtesy of Mom. Well, enough. Nobody was getting any more of me unless they appreciated it. It was time to get back down to New York City, to the place and people I loved.

From: Jessica Locke <jessical@shore.net>
Date: Thu, 13 Jun 2002 23:45:29 -0400
To: Dan Fullerton <danfullerton@yahoo.com>

Hi Dan!

Is it too early to ask whether June 24-27 is a good time to come to the firehouse?

It feels like we are entering another phase of 9/11, and my sense is that I need to do more. I feel very grounded and focused since May 30th. The men know why I'm there, they know what to expect from the sessions, I'm used to the firehouse way of life – this next visit should be very productive at a lot of levels.

I am so looking forward to seeing you all again,

Jessica

From: Dan Fullerton <danfullerton@yahoo.com>
Date: Fri, 14 Jun 2002 12:37:10 -0700 (PDT)
To: Jessica Locke <jessical@shore.net>
Subject: Re: <no subject>

Hi,

Those dates should be fine. It'll be good to see you.

Stay well,
DF

Now that the firefighters knew I was committed, I hoped they would listen to me and trust my judgment concerning their healthcare. Being a firefighter took twenty years of their life – but in terms of stress, lack of sleep, and repeated and complicated injuries – it was more like forty. Many of them worked a second job, usually construction, just to make ends meet. Retirement was likely to be rife with chronic pain. And now, with 9/11 and its aftermath, there was the growing nightmare of evolving and unresolved health issues.

These were not men who would ever ask for help or talk about needing it, and I understood why. Within their special community – their brotherhood – there was no need to ask. To solicit assistance of any kind was against their unwritten code. They had a greater faith. As Dan wrote in his letter: "… if a fireman needed help with his house, guys would just show up."

As I looked back at how consistently my own needs were met in this firehouse, I knew this was true. Not having to ask allowed one to remain strong. Reinforcement was drawn to you instead of weakening yourself by reaching out to someone else.

As such, I realized that the mandated Friday afternoon counseling sessions would only add to their stress. Simply, it went against their tradition. These men responded to action, not talk. While the time I spent with them was brief, it was clear to me that sending in counselors to get them to talk about their feelings was a waste of time and money.

I heard a funny story concerning one of these counseling sessions. One of the firemen told a counselor, "You know, I haven't had sex with my wife since 9/11."

And the counselor, certain he was on the verge of an emotional breakthrough, said, "And how does that make you feel?"

And the guy said, "Fine. I never had sex with my wife before 9/11."

The firefighters would have to rely on each other. And they did. I needed to do more, not because they were asking, but because it was the right thing to do. I would take care of them as they had cared for me. I was learning.

NEW YORK CITY
MONDAY, JUNE 24, 2002
ENGINE 32

Heads turned as I passed by on my way to the firehouse. For the first time in my life I didn't wonder if something was wrong. I didn't question motives. I knew my inner joy – outwardly expressed – was worth a second look. I sensed a greater safety as a woman and, for the first time in my life, truly felt beautiful.

My steps quickened, like in a romantic movie, the closer I got to the firehouse. Today, instead of eagerly ringing the doorbell, I stopped myself. Gently laying my hand on the stone facade of the doorway, I offered up a small prayer of thanks for this extraordinary love that had come into my life.

This particular afternoon I found Ray, Manny, and a guy I had never seen sitting at the dining room table. We said our hellos, and I pulled up a chair. They went back to their discussion and, as usual, I remained quiet. But during a break in their conversation, I spoke confidently.

"Listen, I'm sitting here with a very talented pair of hands, and it really is a terrible waste not to take advantage of them."

I looked directly at Ray, and without waiting for him to speak, I said, "I know, I know, if it ain't broke..." and he joined in with me, "...don't fix it!" That drew a laugh from around the table. Finally, I made Ray smile. It had taken five months.

The new guy said, "I'll try it." He introduced himself as Joel, and once I heard his name, I knew he was the legendary chef of Engine 32. Every firehouse had at least one man who held the title after demonstrating their prowess in the culinary arts. He was very tall, with striking features. It struck me almost immediately that he would have been just as at home in a palace drawing room as he was in the kitchen of Engine 32. Had I still been clinging to a blue collar description of firemen, I would have said he did not belong here. He had a quiet elegance, and his green eyes held a distant look, as though he were living in a future the rest of us had yet to see. At the same time, it was almost like he moved in slow motion. I didn't quite know what to make of him. He asked me how I came to Engine 32, and I delighted in retelling the story. As I spoke, I had this feeling that not only was he listening to every word, he was also allowing each its highest meaning and purpose. I came away from my conversation with Joel feeling as though he had elevated me to greater possibilities within myself.

After his session I asked to watch him cook so I could see the legend in action. I daydreamed about creating a spectacular dinner for these men, but I was still petrified to prepare a meal for others given the horrors of my childhood cooking experiences. It would never be good enough. And here, if I screwed up, they would go hungry. To let them down while they were working would be a terrible thing. Deep inside, I knew I would never offer to cook for them.

Watching Joel cook was like watching ice melt. Painfully slow. Every step was done by hand. He chopped the garlic, he chopped the onions, the mushrooms, everything, even though there was a food processor right there on the counter. At least ten jars of spices were used, and of course, he never bothered to measure anything. I could not imagine spending so much time on a meal. It made me aware of how little time I spent doing basic, nurturing things for myself.

We sat down to eat; it was close to 9 p.m. Breaded chicken with artichoke hearts, roasted red peppers, mozzarella cheese and mushrooms. This was one of the top ten meals of my life. Joel said I ate like I was starving to death.

My letter to Harry, now four months old, was still taped to the front of the refrigerator.

*   *   *

Later on that evening, still upstairs, I heard them calling me. "Jessica, we're being relocated to Engine 33, do you want to come with us?"

I ran downstairs to find out what this meant. Another firehouse was working a major fire, and needed their area covered. They had no idea how long they would be there. I grabbed my belongings since I would be closer to home and could take a cab from there. It felt wonderful to be included, and to have them feel comfortable about taking me along.

As we drove north on the Bowery, I saw a sight that I will never forget. The city lights and the star-filled sky and moon were in a glorious twinkling merger that reminded me of twilight back on the farm in Ohio. There, years ago, before DDT wiped out the bug populations, I saw thousands and thousands of lightning bugs flying over the fields of wheat. It looked just like New York City, but I didn't know it at the time. The beauty of it took my breath away. Ohio and New York City were inexplicably joined in a luminosity that only I had experienced … both then and now.

When we arrived at the Great Jones Street firehouse, the doors were locked, and no one knew the combination. No problem. They raised the ladder to an open window on the third floor and climbed through.

Because it was such a beautiful evening we stayed outside, the engine parked on the apron. Dan, Ray, and I were all sitting on the front bumper. Dan mentioned that this particular firehouse was the most beautiful in the city, and I believed him. Built in 1898, it was a gorgeous building, rising five stories as its turn-of-the-century architectural

155

details adorned it in lovely, effortless symmetry.

I was humbled and grateful to be sitting with these men in the heart of Manhattan on this lovely summer evening surrounded by the intense energy of the city. Engine 33 and Ladder 9 returned about an hour later, and there I stood amidst a collection of glorious huge red ladder trucks and engines, the ground trembling with the vibration of their idling power, the odor of diesel filling the air. It was an extraordinary harmonization of manpower and machines working efficiently and gracefully. It was pure goodness, motivated by goodness.

Before they left, Joel hailed a cab and, despite my protests, paid the driver and sent me on my way.

*   *   *

The cab driver said in an unidentifiable accent, "You work for fire department?"

"No, no, I'm just doing some volunteer work."

He said, "You know, those firemen, they are angels! Angels!"

"Yes, they absolutely are."

"I tell you story about those men. I was in apartment with my child, and there is fire in building. The apartment, it fills with smoke. The firemen come and I am at window with my child, I think we must jump. They say, 'Stay there, we are coming to get you!' And the firemen come into apartment, they come through smoke, and they take us out."

"That's a wonderful story," I said.

"No, no, that is not end of story. The next day they come back! They had break all windows to let out smoke. Next day they come back to fix windows! They know landlord will not do it, so they come back, they fix windows. Those men, they are angels, angels! You must tell them for me."

"Yes," I said, "I will tell them."

TUESDAY, JUNE 25, 2002

The next day, I arrived just as the angels were returning from a high school graduation ceremony in which they were the honored guests. They were wearing their "dress blues" in the 90 degree heat of the day, and were soaked with sweat.

Harry was my first client, and I was so happy to see him. He didn't stop talking, and somewhere in the middle of his monologue he snuck in a sentence about having arranged an afternoon bodywork appointment for me at a local chiropractor's office.

What?! I was dumbfounded, as I had set out to take extra good care of these men on this trip, and here they were making sure I had bodywork, too? I protested that it was my job to care for them, and that I would have a chiropractic treatment when I returned home.

Harry argued back. "Look, it's just an hour. You can take an hour for yourself. You'll be working the rest of the night, so I'll walk you there when my shift is over, okay? Okay? Okay?"

I gave in, knowing it would make him happy if I accepted. I was learning.

I gave one more session and then, still in his dress uniform, Harry walked me up Fulton Street to Broadway. Everyone we passed acknowledged him, but he was humble as they did so, putting his head down, not meaning to draw attention to himself.

Lately I had been suffering from dizzy spells and vertigo. Attributing it to stress and the shock of all that had transpired these past six months, I refused to give it much attention, but was at the tail end of another bout when I arrived in New York. For Harry to have arranged this ... well, as I said before about their code, I didn't have to ask ... it was as though they had anticipated my needs.

We walked a few more blocks and into an office for sports massage and chiropractic. Harry introduced me to the chiropractor, Kathleen, a beautiful Irish woman.

"This is Jessica, the woman I was telling you about who's been taking care of the guys." Kathleen sent me in for a pre-adjustment massage. Harry left, but returned 15 minutes later with, of all things, a large, fresh-squeezed vegetable juice. Through a crack in the door he announced that he would leave it on the counter in the waiting room.

"Did I get it right? Carrots, celery, beet and parsley?"

"Thank you, that's exactly right." I was deeply moved and touched by everything he had done, struggling with the irony of being cared for like this. He told both Kathleen and me to return to the firehouse for dinner that evening. Joel would be cooking a special meal for us.

My head was swimming at this total reversal of caretaking roles. No matter how much I tried to give to these guys, they just quadrupled it back in my face.

Kathleen, as it turned out, had been going to firehouses and giving chiropractic treatments since 9/11. She had to pull back somewhat; she was exhausted and burned out, but told me that any firefighter at the company who needed an adjustment would be treated free of charge, to just send them over. This was a tremendous lift for me, to know there was an additional resource for these men. I told her about my dizziness, and she gave me an adjustment that knocked the last of it right out of me. When she was done, I felt incredibly well. Then I gave her an Alexander session, and afterwards, we walked over to the firehouse together.

Joel had made "surf and turf": Steak and lobster seafood diavolo, with an incredible sauce that I could not stop eating. Between bites and deep thought, its key ingredient dawned on me: Love. The man cooked with love. I would never divulge such a girly-mushy concept to him, but I knew in that moment that I had never cooked with love, only with terror. It made a big difference in the taste.

I gave two more sessions after dinner; five in one day. It was 10 p.m., and I was ready to leave when the call came in that we were being relocated again, this time to Engine 24-Ladder 5 on Sixth and Houston. The guys told me this was much closer to the condo and, again, it would save me money to take a cab from there.

From the moment I stepped down from the rig and into this firehouse, I was deluged by a strong sense of grief. Eleven brothers were lost here on 9/11, and the pain within those walls was so raw I could barely breathe. Their memorial was set against the far wall of the apparatus floor, and I walked over to pay my respects.

The men were gathered together talking as I made my way back over to them. Apparently Harry had been singing my praises and as I approached, one fireman asked, "Hey, would you mind working on my shoulder? I've been having a lot of problems."

"Absolutely."

And then BING BONG, an alarm came in for the ladder company. As they drove out, the firefighter leaned his head out the window and said, "Go home, I'll be alright. It's too late, go home."

I shook my head and emphatically called after him. "No. I'll be here when you get back."

Earlier that day I mentioned to Harry that I had read *Last Man Down*, an account of the 13 men trapped in the stairwell when Tower 2 collapsed. It paid off. Harry came over to inform me that he had talked one of those survivors, Battalion Chief Jeff Ladd, into getting a session with me. Harry introduced us, and spoke of him as "the miracle of 9/11."

E-32 got word that they could return to their quarters. Harry said he felt funny leaving me there, but I assured him I was fine. I wanted to do this and waved good-bye as they rolled away. I was thrilled to be expanding my work into another firehouse under the auspices of Engine 32. It made it far easier than trying to break in on my own, and I was happy to be reaching more men.

This firehouse had a real massage table in their workout room! What a blessing it was to work on a table of proper height and support. Chief Ladd briefly discussed his injuries, and then, fatefully, another alarm came in and off he had to go. He again told me not to wait, to go home, but I refused. Another firefighter stopped in and I offered to work on him until the Chief returned.

I thanked God for Kathleen; the massage and adjustment had given me an extra boost of badly-needed energy. When Chief Ladd finally returned, I worked with him for an hour. Sadly, he was getting me at the end of a long day, and I hoped something of value was coming through my hands. He made no comment, simply thanking me, and I left at one a.m.

Seven sessions in one day.

Giving way to extreme exhaustion I closed my eyes as the bright lights of the city flashed through the windows of the cab. But I had to acknowledge that, once again, I had been treated with respect by men whom I did not know. It was not just Engine 32. This code of honor ran through the entire FDNY.

It was a very comforting realization.

WEDNESDAY, JUNE 26, 2002

The next afternoon Elliot walked into the firehouse, wearing sunglasses, a wildly-patterned shirt, and looking ever so cool. I had not seen him since early May, and the reunion was a joyful one. During his session, he told me he had heard about my "losing it in the firehouse." I must admit, these men talking about me in my absence was unimaginable. I asked him who he heard it from.

"John told me, and then Dan told me." And, he added, "I heard about the cop, too."

"The cop?! You mean to tell me he came to the firehouse after I left?"

Elliot nodded. I was astounded and a little embarrassed. I had to stop and try to take in the fact that even the cop cared enough to ask about me. My feelings had never been important to anyone before. This place – this city – was amazing.

John came into the office. I had not seen him since May, and I immediately apologized for my emotional collapse. He reassured me in a fatherly way that it was fine, and said, "You seem a lot better." I told him that I felt much better and that it had been, in retrospect, a positive thing.

Elliot had not heard the whole story about my giving the poor probie the "massage" over at Engine 7. The look on his face was so funny while I relayed the details, I was laughing harder than he was.

John made a comment about all the noise I was making, and I said, "John, would you rather have me laughing or crying?"

John said, with an absolutely straight face: "When you're here this place falls apart. I can't get no work outta nobody. I run this place like a finely tuned, well-oiled machine, and then you show up and nothing gets done."

A part of me was a bit frightened by the criticism, not sure if he was teasing me, but another part of me was gaining strength in the rightness of my being there, and the belief in the work I was doing.

I countered. "John, this Alexander work will make them better than they were before. They will get ten times more stuff done for having a session, and you should have one, too."

He grumbled something about being absolutely fine and having a back of steel, and that it would hurt my hands to work on him. He got the hell out of the room before I could get him to sit down.

I told Elliot about experiencing all the grief at Engine 24/Ladder 5. He recalled a terrible story about a fire – back in 1994 – when a floor collapsed and three men from that firehouse lost their lives. Elliot said the house never got over it. And now eleven more were taken. How much loss could these men handle? I grew quiet thinking about it. Noticing my descent, Elliot gracefully changed the subject by teasing me again about having a relationship with Dan.

I said, "Look, forget it. It's not going to happen. He has a girlfriend. You guys should set me up on a blind date with a firefighter. Not an officer. Officers get promoted and move all over the city. I want to stay with one firehouse. This firehouse."

Was I really ready for that? No. I needed to stay focused on what I was doing here. But if the day ever came, I knew he would have to be a firefighter.

There was a noticeably more relaxed attitude toward me during this visit. Enough guys had received sessions now that others were convinced I was bringing something good to the firehouse. And so while the acceptance was palpable, I was still afraid to trust it completely. It was the little things that gave me comfort, like getting up from the table and clearing my own plate … no longer being treated like a guest. They still refused to let me chip in on the meals, but I crossed over a major line by clearing the table one evening while they were out on a run. It was risky, I know, a woman treading on their territory, and I worried it would freak them out. Harry, though, took one look at the table and just beamed. "You cleared the table?"

That was that. From then on, whenever the opportunity presented itself, I would clean or wash dishes or help in food preparation. They always graciously protested, but I answered that since they had cooked and fed me, we were even.

Richie, however, wasn't about to let me get away with it. One evening when I offered to help, he literally pushed me down into the recliner.

"Absolutely not. You worked hard today. Here's the paper (places it in my hands). Put your feet up (cranks the footrest), and now you just take it easy and relax. Relax!"

I thought I had just entered the Twilight Zone. Never in the history of the world had I ever heard a man say anything like that to me. He needed to be preserved in a jar in a museum somewhere. Like I said from the first time I met him: If I had ever borne a son, I would have wanted him to be like Richie.

THURSDAY, JUNE 27, 2002

Each firefighter was a unique and special part of the makeup of Engine 32: However, I'm sure it's only natural that some would connect more with my energy than others, and in this regard, Tommy Laughlin stole my heart and soul. I never saw him without a smile on his face. He was in his early thirties, about 5'11" with short brown hair, beautiful hazel eyes and a terrific grin. You just had to smile when you were around him. You could tell he loved being a firefighter; he loved the uniform, loved the radio, and loved using the PA system to broadcast his jokes. He didn't walk so much as he flew. With a natural grace and athletic ability, he could slide down the pole one-handed, and land like a superhero. He didn't have any of the initial shyness toward me that I had encountered with many of the others, partly because he had heard about me long before we finally met. That allowed us to skip six months of "getting to know you" reticence. And the other part: He simply was not shy.

When Tommy was around you had to be on your toes. During this particular visit to Engine 32 the floors seemed to be wet much of the time, and at first I thought they were mopping more frequently. However, I soon learned the cause of this was Tommy, delivering an unexpected bucketful of water over the head of some unsuspecting proby. Of course, Tommy was the comedian who had run in on Dan's session back in May saying "Oooh massage! Can I watch?"

Like a puppy amongst tired old dogs he was forever clowning and keeping things moving in spite of the unspoken grief. I honestly don't know how he did it, but I do know it was intentional. He was just a kid, but something in his eyes showed great intelligence, compassion and determination to do something about 9/11. He was trying to heal it in his own way. Maybe this was our connection; we were both trying to accomplish the same thing. I could only hope to soar so high.

*   *   *

Dan and I were still sitting at the table after dinner, the plates and food having been cleared. Elliot and Tommy were out in the kitchen washing up the pans, and it seemed from all the noise and laughter they were enjoying themselves immensely in the process. Ronnie and Lennie were in the recliners watching the ball game on a television that hung just over Dan's head in the corner. I chose this moment to lean in closer and speak to Dan (under the cover of the rock music and blaring television) about one of the firefighters I had worked with the day before. Although most everyone at Engine 32 had shifted gears through the grief process and kept moving, this particular fireman seemed "frozen" to me. I had not seen him since February, and was careful to note changes in any of them, for better or worse. I decided to mention it to Dan so he could keep an eye on the situation.

As I was talking, Elliot came in and turned off the lights. This was not uncommon, as they often watched television in the dark. Elliot left, but soon returned. Taking the remote control, he shifted the sounds of the radio to soft, romantic music.

It was such a dramatic change from the loud rock music that constantly played, so unusual, that I interrupted my conversation with Dan to interject: "That's different."

Tommy came in with a lit candle, and with a big dopey grin set it on the table between Dan and me. I thought to myself: *A meditation candle. How healthy, how calming. Is it possible I'm becoming a good influence on these guys? Are they actually making a conscious effort to relax?*

No, they weren't. Dan knew exactly what they were up to, these little stagehands changing the scenery into a romantic setting around us, but I was so engrossed with the serious nature of what I had to convey I didn't pick up on it at all.

Dan was getting somewhat antsy, and I wondered if I had offended him in some way. He extricated himself from the conversation. As he was rushing out the door someone said, "Hey, Cap, you should get a session with Jessica," and he responded with some vague comment about having a lot of things he had to do.

Tommy Laughlin came back into the dining room and, wanting to verify their choice of a calmer environment, I asked, "Why did you light that candle?"

He said with a devilish grin, "It looked like you two were having a close moment."

As the realization of what they had done to Dan came over me, I laughed until my stomach hurt. It was priceless. Here was Elliot, fulfilling my wish for a blind date. And then there was me, totally dense and actually thinking they were meditating on a candle and soft music. And of course, when I explained my interpretation of it all, they had an even bigger laugh at my expense.

An alarm came in. It triggered what would become one of my most favorite things; watching these guys transform from easygoing practical jokers into total professionals. As they leaped up I waited eagerly for my cue:

"Are you coming with us?"

We hurried to our places. I was now part of the dance, and I had my own place in the beloved fire engine. Going out on a run and knowing that we were all together, united, one, was living the very best days of my life.

FRIDAY, JUNE 28, 2002

A session with Andy was always guaranteed to be a delight. He was a natural monologue-type comedian, and his stories of his kids' antics would have me gasping for air from laughing so hard. On this particular afternoon, Tommy came in during the session and asked if he could read us a letter he was working on, to get our opinion. A librarian had written to ask what her community should know about 9/11.

He sat down on the bunk next to us and began to read:

Dear Danielle,
What I want you to know the most is something that I've come away with after going to so many masses and memorials. Ok, now, I'm only speaking of the firefighters and not the other almost three thousand innocent people who lost their lives that day. Each of the men I want to speak of whom would lay down his life for you and me was an individual. Each man had his own family, his own Mom and Dad and his own friends. Many of these men were Dads themselves. Most of these men had a wife who now grieves on her own.

So many individual children lost their Dads that day. These kids are not a statistic. They are individuals as well. Each child grieves separately and no child understands why their Daddy doesn't come home anymore.

I cannot group these men together and say, "yeah, we lost 343." It's so much more. This is what I want you and your community to know being so far away. God Bless.

Firefighter, FDNY

Having been taken completely by surprise by the subject (we had not talked much, if at all, about the consequences of 9/11), and the eloquent and honest response, I stiffened up to keep from crying. Andy said it sounded fine. I couldn't say anything. Once I was able to pull myself together, we continued the session. Tommy left. Afterward, I found him downstairs and gave him a big hug.

*   *   *

John O'Brien had completely turned around (or was it me?) since our little "tea party" in the firehouse on May 30th. It was so ironic that the one man I wouldn't want to see me at my most vulnerable was the one that, through my weakness, had connected so strongly. I don't know to what degree that incident had softened all of them towards me, but I now had the distinct impression that most of them wanted me to be happy here, and they would do whatever it took. And they knew it was important to me to make sure every guy had an Alexander Technique lesson.

And so, I was pleased when they pressured John into getting a session on the last night I was there. As we headed upstairs to the office, Tommy got on the PA system and made an announcement to the entire firehouse that John was finally having his "massage." He started singing the melody from *The Stripper,* and then added some rhythm, chanting "boom-chucka-boom-chucka" as we climbed the stairs. I was laughing so hard I had to cling to the railing.

John refused to take off his army boots. He insisted that his back was too hard for my hands. And every two minutes he asked, "Are we done?"

Tommy came upstairs and started working on John along with me, doing a perfect imitation of the Alexander work on John's neck. John muttered and complained while I firmly insisted that this was good for him. Finally an alarm came in. I never saw anyone so grateful to go to a fire. I called after him, "You'll have to come back after your run. We're not finished."

"Oh, we're finished," he said as he bolted for the door. "We're all done."

When they returned from the call, the guys asked, "What did you do to John? We haven't seen him move that fast in years!"

It was a glorious, wonderful, triumphant week.

BOSTON
JULY 8, 2002

From: Jessica Locke <jessical@shore.net>
Date: Mon, 08 Jul 2002 18:42:22 -0400
To: Dan Fullerton <danfullerton@yahoo.com>

Hi, Dan!

What a week, what a heat wave. I don't know how you guys do it. My respect for the work you do just grows and grows with each visit.

We were relocated over to Engine 24 the night that the Captain named George was working. It was late, around 10:00, and I went with them to catch a cab home from there. Harry talked Chief Jeff Ladd into getting a session with me. I had just read *Last Man Down*, the book about those guys who survived the fall of the tower in the stairwell. I liked him very much. We finished at 1:00 a.m. Engine 32 left around midnight. I worked with 3 guys at Engine 24, and there were more waiting. I wished I could have done more.

I wish I could do more.

Something I've noticed in my limited experience so far:
Lieutenants talk less than Firefighters, Captains talk less than
Lieutenants, and Chiefs don't talk at all!

Take care,

Jessica

From: Dan Fullerton <danfullerton@yahoo.com>
Date: Mon, 8 Jul 2002 19:53:15 -0700 (PDT)
To: Jessica Locke <jessical@shore.net>

Hi,

It's good to hear from you as always.

So you got to meet Chief Ladd. There is no one in the Fire
Dept. I have more respect for than him. I haven't read that
book and I don't know what Jeff told you but he is probably
most responsible for getting those guys out of that situation.
Remind me to tell you about that when I see you.

You've been a big help so don't worry about how many guys
you get to work on.

Talk to you soon
DF

PS The guys who talk the least usually know the most.

From: Jessica Locke <jessica1@shore.net>
Date: Wed, 10 Jul 2002 12:07:02 -0400
To: Dan Fullerton <danfullerton@yahoo.com>
Subject: Re: <no subject>

Hi, Dan!

Good to hear from you.

Chief Ladd and I did not talk at all. We didn't have to. Hard to put into words, but I'm sure you know what I mean. You just knew you were in the presence of a great man. I can't wait to hear what you have to say about him and the situation he was in. How long have you known him?

The book was really good in getting a sense of the minute by minute goings on during the attacks and evacuations.

Did you ever write down your account of what happened that day? I would love to read it if you did.

Take care,

Jessica

From: Dan Fullerton <danfullerton@yahoo.com>
Date: Tue, 30 Jul 2002 10:56:56 -0700 (PDT)
To: Jessica Locke <jessica1@shore.net>
Subject: Jeff Ladd

Hi,

Hope all is well with you. We're OK here, just pretty busy.

About Jeff Ladd. In order to get promoted in the FDNY you have to take a very difficult, competitive exam. Your score on this test is combined with your seniority and any meritorious awards you may have received. Then a list is established and members are promoted according to their order on the list. You have to be high enough on the list to be promoted before the list expires and a new test is given. This is usually about every four years. That's the rub. If you don't score high enough to be promoted you have to wait four years for another chance to take the test. Just a little added pressure. The material the test is drawn from is very extensive, and the questions on the exam very technical. It takes years to study for it. To help guys prepare for it (and to make money) a retired chief started a program called Fire Tech. This is where I first met Chief Ladd. He was a Lieutenant at the time and one of the instructors. All officers who study for promotions do so to advance their careers. But with Ladd you just knew he also wanted to be very good at what he did.

After I was promoted I had run into him at a few fires. I always felt comfortable working with him.

On Sept. 11, I was working overtime in Engine 9. Ladd was at that time the Capt. of Ladder 6 which is in the same firehouse. He was on duty also. This firehouse is on Canal Street, about 20 blocks from the Trade Center. One of the members of the company I was in, Engine 9, happened to be standing in front of quarters and witnessed the plane hit the tower.

Both companies responded. What happens next is basically what you saw in the *9/11* film. The members of Engine 9 and myself got out of the north tower seconds before it collapsed. Capt. Ladd and his crew weren't as lucky. They were still in the building when it came down. Somehow, by some billion to one chance, they were in a section of the stairway that was still intact. They had survived but were trapped in the rubble of a 110 story building, a pile of steel and concrete seven stories high.

Me and the members of E-9 tried to get over the initial shock and regroup. When the tower began to fall we had to run to get clear of the debris and were scattered in different directions. By the time I got every body accounted for we had heard Capt. Ladd giving "Maydays" on the radio. There was almost no detectable stress in his voice. No indication of panic. When his message was acknowledged and he had contact with a chief outside he described his situation and location very deliberately. "They had been on the fourth floor of the 'B' stairs."

If anybody was prepared to be in this situation (nobody could REALLY be prepared) it was Ladd. If I had to be in that situation there isn't anyone I would have wanted with me more than him. He kept his head like very few of us could. Once I had all the guys from E-9 together they said, "We have to get to them!" I said, "We will." I had no idea how. We located some tools and equipment. We had dropped what we had fleeing from the collapse. I approached a Battalion Chief near the edge of the pile and told him who we were and that we wanted to get to Ladder 6. There was no way he was letting us onto that pile. I didn't argue, I think he was still in shock. I turned to the guys and said, "We'll find another way in." At this time another Chief approached us and said they had located a passable access to the sub-basement and the base of the "B" stair. He wanted to get a hoseline to this area because there was fire in the stair that had to be put out before we could try to go up to get to Ladder 6. (As it happened this Chief's son was one of the firemen working with me that day. I can't imagine what was going through his head.) We got our pumper working and got a hose in position but this effort had to be aborted. The stair was packed solid with debris and the area we were in was definitely not safe and we were ordered out. When we got back to the street we heard Ladd on the radio. They had found a way out but still had to cross about 100 yards of the pile. This meant climbing over steel beams crossing deep voids. The WTC has six levels underground, some of these voids were over 70 feet deep with aviation fuel-fed fires at the bottom. But there was no other way. I heard Ladd say in a TV interview he told the guys who hesitated, "Your wives and children are on the other side of that pile, keep going." Being able to give that kind of inspiration in that situation takes character few of us can understand.

Well that's sort of an abridged version, I left a lot of stuff out but it gives you a pretty good idea of the type of person Jeff Ladd is.

Stay Well

DF

From: Jessica Locke <jessical@shore.net>
Date: Thu, 01 Aug 2002 14:10:05 -0400
To: Dan Fullerton <danfullerton@yahoo.com>
Subject: Re: Jeff Ladd

Dan, I feel so humbled, like I could not begin to even say anything relevant to your experience of 9/11 and all that came after. Reading that book, and then meeting Jeff Ladd, and then finding out you were an integral part of that situation also ... I know this for a certainty ... that there were no men better equipped to handle what happened. If you had not gone up into those towers, there would be no saving grace to that day. I don't know if you guys know that, being the very heart of it all. There was this horror, met with the most incredible extraordinary force of good men that ever walked the earth. We on the outside got the blessing of who you are. That when we think about that day, we have a choice: To focus on the goodness of men. And find ourselves inspired to move towards greater things in ourselves because of you. I don't know if we'll ever be able to thank you enough. You continue to amaze me. And always will.

I'm hoping to see you all in August, (the men and I discussed coming down for Thanksgiving, by the way, I would love to be there all that week; I hope this is okay with you.) God, listen to me – the men this, the men that. Men men men. That's all I think about now.

You take good care of yourself, see you soon, my love to everyone,

Jessica

From: Jessica Locke <jessical@shore.net>
Date: Sat, 10 Aug 2002 03:36:52 -0400
To: Dan Fullerton <danfullerton@yahoo.com>
Subject:

Hi, Dan!

How are all of you doing with the advent of the anniversary?
I'm trying to feel out a good time to show up again, I was
tentatively looking at the last week of August – something to
lift the men up a little, balance them out, as we approach
September?

Take care,
Jessica

From: Dan Fullerton <danfullerton@yahoo.com>
Date: Mon, 12 Aug 2002 18:32:20 -0700 (PDT)
To: Jessica Locke <jessical@shore.net>
Subject: Re:

Hi,

It might be best if you wait till after Sept. to come down
again. We are trying to be ready to do whatever we need to do
for the families of the guys that were lost. This is going to be
an extremely difficult time for them. More so I think than
they realize. I'd like to keep the firehouse and the guys available
to them any time. We would probably get more benefit from
your work a little later on. We don't want you to get burned out.

I'll keep in touch,
take care
DF

From: Jessica Locke <jessical@shore.net>
Date: Sat, 17 Aug 2002 13:17:34 -0400
To: Dan Fullerton <danfullerton@yahoo.com>
Subject: anniversary

Hi, Dan,

You know that I want to do what is best for all concerned. I am aware that a lot of people's feelings have to be taken into consideration around this coming anniversary.

I am only responding to a request by the men that I come down as soon as I can, and I was trying to avoid getting too close to September 11. John said he was hoping I would come that week, that he would have a video-camera ready and a box of Kleenex for me.

I need to be able to give them a definite answer. Since I last wrote, I've gotten a little more feedback from the men, and am hearing that some are already having difficulties with the anniversary, so I agree with you that we should wait until after the 11th. The week of the 16th would work for me.

I don't like to leave the men hanging too much longer after the 11th, knowing how much they will have given out to the families and to the public.

Dan, I have made this investment of time so that I could be there for them when it got rough, and that they would be able to trust me. We all knew this time was coming. My work is not recreational; but if I had made it look serious they'd have never let me in. I wish I could tell you what goes on in the individual sessions so that you could better understand my assessment, but I can't. It leaves me powerless to have any influence on your assessment.

Know that I have their and your best interests at heart,

Jessica

P.S. I'll be sending you the CD of the memorial music I wrote for the firemen, I'd be honored to have you listen to it.

From: Dan Fullerton <danfullerton@yahoo.com>
Date: Sun, 18 Aug 2002 19:19:08 -0700 (PDT)
To: Jessica Locke <jessical@shore.net>
Subject: Re: anniversary

Hi,

The week of the 16th should be fine. I'll be on vacation around then, but I will probably be back in time to catch you before you leave. I'll be able to relax a little easier knowing you will be there keeping an eye on them. You know what they mean to me.

If you can, send us an extra copy or two of your CD. I'm anxious to listen to it. Talk to you soon.
DF

From: Jessica Locke <jessical@shore.net>
Date: Tue, 27 Aug 2002 08:40:58 -0400
To: Dan Fullerton <danfullerton@yahoo.com>
Subject: program notes

Hi, Dan,

I just heard from Andy that the CD's arrived, I sent you 3
copies. If for any reason you would like more, let me know,
I'll be happy to make them up for you.

My idea for this memorial piece was to use the names of the
firemen as the "words". I wanted to get across to people that
"343" is more than a number. So all the firemen's names are
sung during the course of the piece. You'll understand what I
mean when you hear it. It ended up being extremely difficult
to do, and I am not sure any choir would undertake it. So I
think this is the only performance of it that you will ever hear.

Take care,
Jessica

From: Dan Fullerton <danfullerton@yahoo.com>
Date: Tue, 10 Sep 2002 08:28:40 -0700 (PDT)
To: Jessica Locke <jessical@shore.net>
Subject: Re: <no subject>

Hi,

I'll be at the firehouse on the 11th just to check on the guys working. At some point that day I'll go over to the WTC site to pay my respects.

It may be a difficult day so be strong.

I'll see you when I get back.

DF

SEPTEMBER 2002

As the first anniversary of September 11th loomed, I felt myself closing down with a sense of dread and sorrow.

I was helpless to do anything for the firefighters; they had to be going through a terrible, personal hell. Had I done the right thing, to withdraw at this time? But they were shutting down in order to get through it, and needed to close ranks.

It was going to be hard for me to get through this on my own, but I could not look to them for support now, even if I needed it. The invisible line between them and me had to be respected. As their caretaker I needed to maintain my objectivity in order to keep a keen perspective on their requirements. The big picture of what they had been through was bad and, as resilient as they were, I knew even worse times were ahead. In retrospect, I had reached so few men, and felt guilty that perhaps I had not done enough.

A week before the anniversary, it dawned on me that I should have done something with the memorial piece. These men deserved a singular tribute, and in a desperate last-minute move I sent a copy of the CD to WCRB, Boston's principal classical station. On the outside of the envelope I wrote: MUSIC FOR SEPTEMBER 11.

Most stations plan their programs at least a month in advance. In the case of 9/11, they had a pre-recorded program ready to go. I was too late.

I woke up early that morning, dreading the endless replays of the tragedy that would go on all day. The telephone rang at 8:30 a.m. I glanced at the caller I.D. Unknown. I answered the phone, and a voice said, "Is this Jessica Locke?"

"Yes."

"This is Laura Carlo's assistant from WCRB. She asked me to call you to turn on your radio; we are going to play your piece in ten minutes."

Oh, my God. It was unbelievable. Oh, my God. I called Dawn, telling her to listen.

And at 8:40 a.m., Laura Carlo broke into the pre-recorded program with a live insert to read from the letter I had written to her.

"... this is a painting of many funerals, many tears, and the continued strength the firefighters showed through the months after 9/11 in the cleanup of the World Trade Center site."

And then she played my piece.

The morning was completely transformed from a singular grief into having my wish to honor the firefighters accomplished with the playing of this music for all to hear. How many people were listening that morning, remembering a year ago, privileged to stand in the light and grace of these extraordinary men?

And then, the radio program continued with a piece by Johann Sebastian Bach. There I was, a musician whose composition was followed by Bach. Again, in my desire to give to the firefighters, they had given me back new stature in my own life.

My energy shifted and I felt safe, as though I was back at Engine 32. The grief was somehow detained outside of me. I would get through this day strong and secure.

I wondered if I should have sent them flowers ... or a note. I wasn't sure. Being in such a relationship – where love flowed both ways – was new ground for me. I didn't know what was appropriate.

I decided to send an email:

From: Jessica Locke <jessical@shore.net>
Date: Wednesday, September 11, 2002 12:01 p.m.
To: Engine 32 (info@fdnyengine32.org)
Subject:

To all of you:

Throughout this year, all of your names stand out as the finest men I have ever had the fortune to meet and know. I thank you for helping me to feel safe. I thank you for allowing me to do something at a time when I felt so powerless. I thank you for showing me how to act and be in these extraordinary times.

I love you all so much.

Jessica

MANHATTAN
SUNDAY, SEPTEMBER 15, 2002

As much as I was certain of the "entity" – the voice that brought me to NYC in the first place – I was hesitant to trust that I was wanted at Engine 32. The men were polite and gracious while tolerating countless tourists and guests, and it occurred to me that perhaps they were simply tolerating me, too. I had a unique ability to delude myself where men were concerned. I wanted this so badly, it was possible I had overlooked whether they wanted it. As long as I needed them I knew I could stay, but now I had to know they needed me, too.

I hated these *girl* feelings; the universal need for "commitment." Just mentioning that word freaked out every man I had ever been with. But the issue here was: If the firemen didn't really want the bodywork and had simply tolerated me in order to help me feel safe and loved, I was going to move on. The true act of love would be to let go. And I loved these men more than anyone or anything in my entire life.

I knew they would be honest, blunt, and straightforward. If I had the courage to ask, I would hear the truth. The entity said "yes." Every time I asked, the answer was "yes." But I was tired of trusting the ether. I wanted a reality-based version.

On Sunday, September 15th I arrived in the city and called the firehouse to see who would be working the next day. I found that there would be training all day and no one would be at the firehouse until 5:00 p.m.

Not a good start. A good portion of the day would be wasted. Was it a sign that I was coming to the end of my relationship with them? In spite of my initial reaction, I had come to trust that whatever happened in NYC happened for a reason. Now I had a free day. What did I want to do?

I wanted to work with firefighters.

I called Engine 24-Ladder 5. I reminded them of my visit back in June and that I had worked with Chief Ladd. I added that my afternoon was free, and asked if he wanted me to stop by.

The firefighter put me on hold, and came back and said, very directly, "Yes, he would like to see you. Can you come at 2:00?"

Well, somebody wanted me. I asked for subway directions and arrived at 1:40 or so. No one was there. Every firehouse in the city had memorials outside quarters, and here there was a vast array of flowers and candles. Eleven men had perished from this house, and the grief was as intense as when I first visited. Elliot's story about the three others who died back in 1994 still burned in my memory: *The house never recovered from that, either.*

I walked over to a nearby flower shop and purchased a lavender rose, and while returning, saw a red fire department minivan drive up. Chief Ladd got out. We smiled shyly at each other, and I made a "just a minute" motion with my hand as I hurried over to place my rose in one of the many vases out front. Walking into the deserted apparatus area, we exchanged pleasantries.

"I am so glad for this chance to redeem myself. I felt so bad that you got me at the end of my energy that night."

He said, "That was the best massage I ever got in my life."

"Well, if that was the best massage you ever got in your life, then you are in for a treat because you are my first session this week and I've got all my energy!"

It was a good session. I told him it would be very beneficial to have two lessons in a row if possible. Was he around this week? He said he

186

would be working again on Thursday. I thought that if Engine 32 didn't want me anymore, I could start cultivating a relationship with this firehouse.

Afterward, I walked back through the apparatus floor toward the street. Two firefighters stood there, clearly hoping I would offer to work with them. I knew they would never ask. It broke my heart to have to tell them I was expected at Engine 32 and had to leave. They shrugged it off, but I left feeling as though I had let them down.

I took the subway to Fulton Street. As I turned the corner, I saw John leaning against the open garage doorway. I did this little dance, so excited to see him, and we pressed our faces together in greeting. "What've you been up to, little lady?" he said.

"I was doing Alexander work at Engine 24-Ladder 5," I replied, proud that I had added another firehouse to my collection of adoptees. "It's really sad over there, John. They haven't done as well as you guys have here. Maybe it's time for me to move on and take care of their house."

A determined look crossed his face. "You stay right here."

I almost fell over. The senior man who didn't want any woman's jewelry anywhere in his firehouse was telling me – no, *ordering* me – to stay on at Engine 32? I wanted to hear more, so I dragged it out a little.

"Well, you know, I made this promise to come down here for a year, but I haven't been sure you wanted me, and now I need a commitment, John. I need to get married; I need to know you guys want me."

There, I had said it aloud. I wanted a commitment. Elliot had approached us during this conversation, and he nodded his head in agreement with John, saying, "You stay right here at Engine 32."

I resisted the desire to whack myself on the side of the head to be sure I wasn't hearing things.

It was so easy! Was it possible that when love is real, commitment isn't an effort, it's just a logical conclusion? The concept was profound to me.

Later on that evening, while the guys were preparing the meal, Ray said in his heavy Brooklyn accent, "So, Jess. We've been cooking dinner for you all year. When you gonna cook dinner for us?"

This was the first time Ray had ever asked anything of me. In fact, I think it was the first time he had spoken an entire sentence to me except "if it ain't broke, don't fix it." I recognized this as an opportunity to connect with him on another level.

I reacted quickly, "Okay. I'll do it. When are you working again?" He said Thursday night. I accepted the challenge confidently. "Thursday night, then. I'm cooking."

John poked me with his finger and grumbled, "Don't you go experimentin' on us."

That comment made me recall that I only cooked with terror.

Just then, in a supremely unique moment, I realized I had never experienced fear in this firehouse. They had let me do only what I felt comfortable doing, and so far I felt perfectly safe. It was almost as though Ray instinctively homed in on my greatest weakness, and was pushing my boundaries in order to make me stronger.

This would be the most important meal I would ever cook. The old me freaked out, but the newer me – the safer self – knew I could handle it. Joel and the others cooked with a slowness and careful handling that spoke of love. The meal would be okay because I loved them all. I had two days to plan the menu. I could do this. I would cook with *love*.

TUESDAY, SEPTEMBER 17, 2002

On Tuesday, we took off in the rig one more time for the evening grocery run. The joy of riding in the fire engine never lessened. That huge, lumbering, magnificent animal took care of us, protecting and shielding us from the outside world. As we drove beneath the Brooklyn Bridge, the sea air blew in against the heat of the engine. A huge full moon rose in the night sky, its reflection shimmering in the river. I glanced quickly at the faces of the men. Nobody spoke. The silence held our thoughts.

The grocery store parking lot was filled with rigs from Engine 4 and Engine 7. We got out while Elliot stayed with the engine, as usual, and we made our way through the automatic doors into the store.

The guys grabbed a cart, and after some discussion of what they needed to buy, they split up and set off in different directions. Manny went to pick up some bread, and out of the corner of my eye I saw that the men from Engine 4 had gathered around him. I assumed he knew them all quite well, and didn't think anything of it. As we made our way around the store, I began to notice that the men from Engine 4 would occasionally glance over at me.

Manny returned, a big smile on his face. "Did you see how those guys swarmed around me? They wanted to know who you were."

"What did you tell them?"

"I told them you were a massage therapist, and that you came down from Boston to work on us. So they said, 'Hey, send her over to our

house when she's done with you, okay?' But I said, 'Nope. She's all *ours*'."
Manny emphasized these words with childlike glee. "I said to them, 'she belongs to *us*'." Manny, usually so quiet, was now quite animated and enjoying every bit of this. Competition between firehouses was fun and fierce. It was now a matter of one-upmanship, and Engine 32 had their own exclusive massage therapist.

As we left the store, Engine 4 followed close behind. One of the men got down on his knees on the pavement and held up his hands in prayer fashion to me, saying, "Pleeeease, just five minutes on my rhomboids, just five minutes?" Elliot looked down from the driver's seat with one eyebrow raised, not knowing what the hell was going on. I was giggling, but knew I couldn't stop for this guy. *I belonged to Engine 32.* It was confirmed and it was real.

It was a commitment.

*   *   *

Harry's car had broken down in front of the firehouse a week prior to my visit. Taking up a valuable parking space on the always overcrowded street was starting to become an issue. More than once I overheard the men asking Harry when his car would be towed.

On this particular evening, both Ray and Harry were walking me up the street to catch a cab. As we approached Harry's disabled car, Ray noticed movement within the dark interior: Someone was inside, head beneath the dashboard, attempting to hotwire it.

Ray said, "Harry, somebody's in your car."

Harry quickly headed toward the vehicle, but Ray reached out a hand to stop him.

"Let's see if he can get it started first."

WEDNESDAY, SEPTEMBER 18, 2002

It was early afternoon. I was taking a break from the Alexander work and had gone downstairs to the kitchen to get a glass of water. As I entered the dining area, I found Mark teaching Lennie – one of the probies – how to make a rope harness. I quietly took a seat to watch and learn; taking in yet another aspect of these men and their work.

Mark was an excellent teacher. He spoke slowly and quietly as he demonstrated how to tie the complicated knot and step into the harness, then how to secure the rope around his waist. The rope would be tied to a chimney on a rooftop and the firefighter would be lowered down for the rescue. As the lesson continued, Mark referred to the technique as the "Paddy Brown rope harness," and it was clear the name commanded respect and admiration, as the proby seemed to know of him, too.

I studied Lennie's face as he absorbed the information. Here was a guy that, to the outside world, was so tough. Watching him in this open and vulnerable role as student was special, yet there was an intimacy about it that made me a bit uncomfortable; the making of a firefighter is humbling. Mark ended the lesson by showing Lennie how to repack the rope into its container, an important detail. If the harness was needed in an emergency, it must not emerge in a tangle.

Throughout the lesson, I acquired enough information to understand that a fireman named Paddy Brown had lowered some firemen from a roof in this harness; he was able to save some people trapped in an apartment fire out of reach of the ladders.

I envisioned this fireman having accomplished these heroics in another era – the 1920's perhaps. He became a hero for saving those lives with the harness, and throughout the century his name had become one of the glorious legends of the FDNY.

As soon as I could, I searched the internet to learn more about him and his story.

Captain Patrick J. "Paddy" Brown was, indeed, a legend. He was the most decorated firefighter in the history of the New York City Fire Department.

Paddy Brown died on September 11, 2001.

THURSDAY, SEPTEMBER 19, 2002

This was going to be a huge day for me. I was going to have to do the grocery shopping in the morning since Tommy Laughlin wanted me to be at the house for lunch. He was making his special stir-fry. They would be gone most of the afternoon for building inspections, so I was going to Engine 24 to work on Chief Ladd. Then I would come back and make the dinner.

I decided to make Chicken on the Spit, a recipe by a chef named Luigi. I had made this only once before, back in 1988, and quite by accident it was terrific. I called my brother back in Boston to get the recipe. It required fresh pasta, and I searched the Yellow Pages to find it … somewhere on Ninth Avenue.

I continued on my way to the grocery store. Down in the subway I sat on an empty bench, the special moments of the previous days passing through my mind.

A gracious middle-aged man walked up to me and, in an unknown accent asked, "Excuse me, but I have to ask you, what are you thinking that could put such a beautiful smile on your face?"

Gesturing with my bags of pasta I replied, "I'm going to cook dinner for some firefighters."

"That's very nice of you," he said. "Where are you coming from?"

"Boston."

His eyebrows went up. "You come all the way from Boston to cook for them?"

"No, no, I'm here in the city for the whole week. I've been volunteering at their firehouse."

The next thing I know, tears are pouring down his face. "I can't believe you are doing this for our firefighters," he said, pulling a handkerchief out of his pocket and dabbing at his eyes. I gave him a CD of the firefighter memorial piece (which I always carried with me). The gesture made him cry harder still.

On to Whole Foods Market. I had no idea how much the men would eat. How much lettuce for a salad? Two heads, or three? I agonized and overbought. I grabbed a taxi and as I pulled up to Engine 32 Tommy Laughlin was standing out front.

"You're late," he said. "You missed my lunch."

I apologized profusely, and he assured me he had saved my share. Other guys grabbed the grocery bags from my hands and brought them to the kitchen.

I went in, and the stir-fry was waiting on the table for me. God bless him, I was starving. It was one o'clock and as the men left for the building inspections, they asked if I was going to be alright. I told them I was going over to Engine 24 to do some Alexander work and would be back when they were finished.

It all went off like clockwork, and we met back at E-32 at 4:00.

This was it, my big moment. I squared my shoulders and went into the kitchen to prepare the meal. Yes, I was going to cook with love. I was ready.

Slowly and carefully I cut the chicken into squares for the skewers, removing the tendons and fat. I paid attention to the smallest detail. So far, so good.

The recipe called for the skewers to be dusted in flour. I looked all over the kitchen. Where was the flour? Ah. I spotted a big round barrel next to the stove. Firefighters need to have lots of flour. I measured out a cup and poured it onto a plate. It was grainy. I smelled it. It was detergent.

John came in. "How's it goin' in here, little lady? You doin' all right?"

Standing so that he couldn't see the plate of detergent, I asked, "Where do you keep the flour, John? I can't find it anywhere."

He pointed to a set of stainless steel canisters right in front of my face on the shelf above the counter, the first of which was clearly labeled *Flour*. "Flour. Sugar. Salt," he said, touching each one like I was Helen Keller. Then he noticed the box of mushrooms on the counter. "I can't eat mushrooms, I'm allergic. You'll have to leave them out."

I assured him that I would make two separate sauces.

He started out, and then turned back, "Now are you okay in here? Is there anything else?"

"Oh, I need to know how to turn on the stove." I was pretty sure it was the kind where you had to light a match while turning on the gas, since it was nearly antique. I had never seen any of the guys turn on the stove while I was in the kitchen; it had always just been on.

John said, "Oh, it's very, very difficult." He came over to the stove and slowly turned the knob and the flame shot up. He gave me a look and left. He must have wondered if he was going to get any dinner at all that night.

Mark came in. "Can I help you with anything?"

"I think I'm okay, Mark, thanks."

"I'll let you in on a little secret for cooking in the firehouse: 'Feed 'em late and feed 'em alot.' If you keep them waiting 'til 10:00 it won't matter if you serve them cardboard, they'll eat anything." He got me laughing and then he said, more kindly, "You know, you don't have to worry. If it's bad nobody will say anything. They appreciate the fact that you cooked. But if it's really bad, they'll just say, 'pass the hot sauce.'"

He left.

A few minutes later Ray came in. "How's it going, Jess?"

"Good, good." Did I look confident as I skewered the chicken with fresh sage leaves and prosciutto?

"You need any help? I can send Lennie down."

"No, I think I'm okay."

"It's not too late, we can still go out for the meal."

"I think I'm good, I can handle it." I thought it was sweet that he was trying to let me off the hook.

The recipe called for barbequing the skewers over a bed of chopped onions, celery, and mushrooms, so that the flavor dripped down into the vegetables. I never did understand how Luigi did this, as the recipe was not terribly clear. The sauté tended to burn where the chicken didn't cover it, but I stayed on top of it with a spoon and managed to get the skewers cooked just right, a little underdone in the center. While they continued to cook at a much lower temperature in the oven, I started the tomato sauce.

John wandered back into the kitchen, eyeing the mushrooms on the counter. "Are you sure there aren't going to be any mushrooms in this?"

Oh Lord, the mushrooms. There was no way to execute two versions of this meal. I was running out of time. I assured John there would be no mushrooms.

He left. I don't know who was more worried, him or me.

Creeping into the process came the anxiety. I had been unable to find the brand of tomatoes I liked, so I chose some organic tomatoes with basil leaves in it. This is the part of cooking I didn't understand; flavors. Luigi did not mention basil in his recipe.

Lennie, the new probie, came down and offered to help. (I'm sure Ray told him to.) With panic setting in, I accepted his offer. He washed the lettuce and put it in the salad spinner, then chopped up the rest of the salad ingredients. As he worked, he said, "You know, it's a real honor for the men to trust you with their meal. It means you're *in*."

Was he teasing me? Forty years of women's liberation, and it was an honor to cook for men? A year later I would learn he meant it. It was a tradition, a rite of passage, that when a probie came to the end of his first year he had to buy and cook the meal. It was a sign of respect and acceptance that Ray had asked me as well.

And now I was about to blow it big time with the tomato sauce, which was a disaster. The recipe called for dry white wine. Not knowing anything about wine, and having always had a particular dislike for dry wine, I instead bought the sweetest wine I could find, not thinking it would make any difference. In any event, the sauce tasted like turpentine. Was it too much onion? The absence of mushrooms? I didn't know. I resorted to standard operating procedures from Ohio: When in doubt, add sugar. I poured in a cupful, but the turpentine flavor remained, only sweeter. I added more onions. I added more sugar.

The panic rose up as it always did when I cooked for people; it was just never good enough. However, this time I sensed the calm voice of the entity. "It's all right. Half the meal will be okay, just tell them the sauce is bad. It's going to be what it's going to be."

There was no saving the sauce; it was terrible. I had water boiling for the fresh pasta, a huge stockpot. I asked Lennie if he would pour it into a colander when the time came, as I would not be able to lift it. He agreed.

Being fresh pasta, it would take all of 60 seconds to cook, so as soon as I dropped it in, I ran out to the apparatus floor to get Lennie. "Okay, I need you!"

BING BONG! "ENGINE!"

I will never forget the look on Lennie's face as he saw the look on mine.

As he leaped for his bunker gear, I raced to the kitchen and started bailing water out of the gigantic pot. Steam poured up from the sink, and sweat poured down my face. I grabbed two potholders and managed to pour out the noodles into the colander. It was too late. The pasta was mush.

I decided to take a chance that they would have a short run. I finished up the recipe, dressing the pasta with butter and parmesan cheese. The recipe called for ladling the tomato sauce over the pasta and then laying the skewers on the top, but I didn't dare let that toxic tomato sauce ruin the rest.

The table was set, so I laid the skewers on their own plate, and served the pasta separately. I put the sauce in its own bowl. I completed the setting with salad and French bread. When the men returned, dinner was on the table. This was it. The anticipation was killing me.

I announced that the sauce was a disaster and that they could try it at their own risk. "Well, it looks good," Ray said encouragingly. They all dug in, and after a first bite, they all said in unison, "PASS THE HOT SAUCE!"

Lennie, God bless him, said he liked the toxic brew and had two helpings. I was amazed. Everyone else said the chicken and everything else was really good.

John said, "I called Pace Hospital Emergency Room and told them they could stand down, we are all okay."

I had a feeling that I had just been given a great honor. No college degree, no film festival award, nothing else I had ever accomplished felt like the privilege of having cooked a meal for these men. It was one of the happiest days of my life.

Ray asked me what I had done that day and I told him about going over to Engine 24.

"Oh, so you're abandoning us now for Engine 24?"

This was it, the point of no return.

I needed to tell them the truth. I needed to let them know that it was I who needed them … that the bodywork was a pretext, a cover-up for my own healing. And now, in this very moment, I had to know if they needed me, too. My little speech was all prepared.

"Well, you know, I've been coming down here unannounced, uninvited, to do this bodywork that nobody wanted …"

Ray interrupted me, "You're just figuring this out now?"

The men erupted into the loudest laughter I had ever heard. They were choking, laughing themselves silly. The look on my face told it all. They had known all along that I was using the bodywork as an excuse to stay. I turned beet red. But it served to confirm that the voice of the "entity" was real. Everything matched up. They knew everything. I had been totally transparent.

"Sharp as a tack, this one!" John chimed in.

After they recovered, I announced that we were coming to the end of my one-year commitment, and that maybe it was time for me to move on to another firehouse.

Ray sounded indignant. "So you're just gonna leave us?"

I couldn't tell whether he was serious or not. "Well, you are all smiling now, but you weren't smiling when I first came down here."

Mark said quietly, "We're smiling on the outside but we're dying on the inside."

Ray said, "Post traumatic stress goes on for years. You're gonna walk out on us now?"

Then Ronnie, who had never said more than two words to me, added six more: "It hasn't even hit us yet."

There was a gentle pause as we all stopped to consider our places in the history of 9/11.

"Allright, then, I'll stay. I never ever want to leave Engine 32. I'll renew my contract for another five years. I just need to know that you want me here. So many people have been landing on your doorstep, and I don't want to be another one taking from you."

Mark replied, "If we didn't want you here, you wouldn't be here. End of story."

It wasn't the most romantic way of putting it, but in the black and white world of men it was just that straightforward. *I wouldn't be here if they didn't want me here.*

For the first time ever, my entire being simply relaxed.

OCTOBER 2002

The groundwork was set. It was time for me to do the best work of my life.

Prior to 9/11, I had always believed the definition of "commitment" was the ability to endure a relationship even if you were being treated badly. But with Engine 32, I felt a state of grace in my willingness to commit myself to them, no matter where it led. For the first time, my life made sense. Everything I had previously learned – or suffered – prepared me for this moment. Engine 32 had earned my respect and love, and I was never going to leave them.

However, I was not feeling well. Fatigue, vertigo and dizziness were becoming more of a problem, with each new bout seemingly worse and longer lasting. The divide in my head was gone, but the impression it left continued to ache. Of course, I worried I had a brain tumor like my mother, but my close encounters with the limited vision of Western medicine would not allow me to place myself in their hands. If I was dying, I preferred it to be on my own terms.

Unable to consider another trip to NYC under such circumstances, I finally went to see a homeopathic[1] medical doctor. The description of

---

[1]A system of medical treatment based on the use of minute quantities of remedies that in massive doses produce effects similar to those of the disease being treated.

my symptoms confounded him; he had never heard vertigo described like this, which led me to believe it was most likely, and not surprisingly, stress related. He prescribed a remedy he had never given to anyone before. It was a preparation for cholera. One of its chilling references: *child hiding in the closet, screaming.*

I recoiled. Never did I scream as a child, I insisted. To which he replied, "You were screaming. You just did it silently."

The remedy worked. If it was a brain tumor I would not be feeling better, and that was reassuring. Yet, in order to go to NYC and work, I needed some energy – certainly more than I had. I chafed at the slowness of my recovery.

I also worried about the firefighters. Now that the first anniversary had come and gone, so too had the great levels of glory and compassion. The men would call it: *From heroes to zeroes.*

In a book published just after the anniversary of 9/11, a vicious accusation was leveled against the firefighters: It said that the crew of Ladder 4 had stolen a stack of jeans from a store located in the concourse beneath the World Trade Center. They supposedly brought them out and stored them in their fire truck before going into the blazing building. The author claimed that the jeans were discovered when the buried truck was dug out of the rubble. It was in newspapers across the country. It made me sick.

How ludicrous that during the biggest fire of their careers, in the presence of thousands of people, men who took great pride in their profession and in their reputation would run downstairs to the underground mall, steal a random pile of jeans, bring them outside to their truck, grab the jaws of life and proceed to run toward the disaster of the century in order to rescue people, giving their lives in the process.

I know these men. I know this story is untrue, and I will never believe otherwise.

I was fortunate to speak with a firefighter who worked on the site every day and was present when that rig was dug out of the rubble. He told me there were no jeans in the truck. I report this here and now for the sake of an accurate history. Yet, my own personal experience regarding the standards these men live by is proof enough for me.[2]

The fact that someone made such an accusation despite all these men had been through, had lost, and had given, was an abomination. With the publication of the book, the media pounced on the story as though it was God's word. The FDNY wisely chose to issue only one comment, and I prayed that the lack of controversy would stall sales of the book.

Then, add to this the long list of other insults: Arrests at the WTC site for trying to recover the remains of their brothers and doing right by their families; forced firehouse closings; becoming tour guides for Ground Zero; and, providing public relations for the city while trying to do their jobs. There had to be a breaking point. I didn't know how much more they could take.

Something had to be done.

*   *   *

The muffler gave out on my car. I had one of those Midas lifetime guarantees; this would be muffler No. 3. However, the car was so old, they didn't have the muffler in stock. It took six weeks to find one buried in a warehouse down in Texas. Meanwhile, I set off car alarms everywhere I went through sheer noise and vibration. During a spell of vertigo, driving anywhere was a nightmare. It was a huge relief when the muffler was finally replaced. Again, the garage mechanic warned me about the axle: It could go at any time and it was dangerous to drive

---

[2]Verified in Charles Pellegrino's *Ghosts of Vesuvius,* a documentary aired on the History Channel in 2006. The excavation of the truck was filmed from start to finish. There were no jeans in the truck.

the car at all. There was nothing I could do. I didn't have the money to buy a new car. Or a used one either.

I drove only when necessary, avoided the Boston potholes like a video game warrior, and prayed that the car would hold out just a little longer. It did. God bless my Audi.

NOVEMBER 2002

While continuing to recoup energy and health, I had this crazy idea to enter *Reading of the Names* for the Pulitzer Prize in Music. With major attention brought to the piece (should it win), I would have enough money to continue my work in New York for the long haul. However, there was one catch: In order to be eligible, the composition would need to have a world premiere performance.

I sent the recording to my brother, an orchestral and choral conductor at Kenyon College. You know how family is. I had no idea what he might think about the piece. But I received an email back almost immediately. He loved it; told me my work was beautiful … overwhelming. He would prepare and conduct the Knox County Symphony Orchestra and a small chorus from Kenyon College. They would perform a world premiere on February 8, 2003.

I was, quite frankly, stunned. *Yes* was a word that continued to surface in everything I did concerning the firefighters.

Shortly thereafter, my sister informed me that the Masterworks Chorale in Toledo, Ohio was giving a *Tribute to Heroes* concert. I sent them a CD as well: They would perform the piece as part of the tribute on March 8, 2003.

I listed the piece on an internet site called *Orchestralist*. The fellow running the site scheduled four performances with an orchestra in California for April 2003.

A conductor from Michigan found me on the site and scheduled two performances for November 2003.

Ron Della Chiesa, a popular radio announcer in Boston, agreed to play the piece on his show, *Music America.*

That piece – which I never thought anyone would perform – now became a full-time job. Orchestra and choral parts had to be created and printed for the musicians and singers. It took three weeks just to complete the chorus parts, time consuming since each name had its own rhythm, and so had to be placed in a particular order that made musical sense.

Program notes would have to be written. I asked Tommy Laughlin if I could include the beautiful letter he had shared with me last summer. He wrote back:

11/23/02 12:20 AM, tlaughlin@cs.com at tlaughlin@cs.com wrote:

Jessica Locke,
Sorry for the late reply, i haven't been spending much time on my computer lately. i hope you are doing well. it is getting cold here so it must be cold where you are. stay warm. sure you can use my letter. anonymous would be nice. do what you want. you have my blessing. see you in dec??
tom

From: Jessica Locke <jessical@shore.net
Date: Wed, 27 Nov 2002 02:01:16 -0400
To: <tlaughlin@cs.com
Subject: Re: program notes

Hi Tom,

Thank you so much. Your letter is going to be a powerful addition. I heard from my brother that the orchestra that rehearsed my piece (at the first rehearsal) fell into a strange silence afterwards. He said it was hard to conduct (from sadness.) I'm glad people are still willing to remember.

Yes, you will see me in December! I emailed Dan yesterday, requesting his blessing on a visit to Engine 32. The dates right now are Dec. 13-19. I hope you will be there! If you should run into Elliot for any reason, would you please let him know I would love to see him?

Take care, Happy Thanksgiving to you and yours!

Jessica

From: Dan Fullerton <danfullerton@yahoo.com
Date: Sat, 30 Nov 2002 14:29:59 -0800 (PST)
To: Jessica Locke <jessical@shore.net
Subject: visit

Dear Jessica,

Since the last time you were down to see us the Battalion Commander (my boss) held a meeting of all the company commanders. One of the issues he addressed was the return to normal Fire Dept. routines. In the months after September 11 things in the Fire Dept. were anything but routine. Many longstanding rules were not enforced if not formally suspended. Discipline had become somewhat lax. This was understandable and to a point excused in light of the stress the members had been under: working at the recovery, daily funerals, extended overtime etc. This, he assured us was about to change. There has been extensive restructuring in the upper management of the Dept. We could expect unannounced visits from Staff Chiefs (HIS bosses) any time. These are on the order of what might be called "surprise inspections". The purpose is to ensure a return to the discipline that is essential in an organization such as this one. The apparatus, equipment and quarters must be in order. Drills and instruction must be conducted. They are looking to run a Tighter Ship. Our regulations are clear about visitors in quarters. Generally no visitors are permitted in quarters after 10PM and not above the apparatus floor. (The chief made special reference to this in his meeting.)

An inordinate number of senior members have retired in the past year. This leaves us with many newer members with a lot to learn. The Dept. is in a rebuilding mode at a critical time. This is a city still at great risk with a Fire Dept. of largely inexperienced and untested members. While on duty they need to be focused with as few distractions as possible.

I am aware of and share your deep concern for these men. The Fire Dept. has, however, assigned a counselor/psychologist to every firehouse that lost members. I have met with the one assigned to Engine 32 several times. I found her to be a dedicated professional. She comes to the firehouse with an associate every week at a predetermined time so as not to interfere with our scheduled activities. They are also available at their office for any member who decides he has issues he needs to address one to one.

I want you to know that you are welcome at the firehouse any time but for you to continue to conduct therapy sessions in quarters would no longer be appropriate. This decision is not an easy one but making difficult choices is sometimes part of my job. Please understand that it is made with the welfare and professionalism necessary for the safety of these men paramount. Please think about this before you make plans to come down and let me know what you decide.

I hope to hear from you soon,
DF

The sense of loss was so great I could not comprehend, nor withstand it.

My thoughts ricocheted like bullets in my brain, gaining a momentum and speed that would defy a rational conclusion. *Return to normal?* Were they all crazy? Could we turn a switch and get back to life prior to 9/11? Did they not get just how bad this was?

Within my mind I desperately searched to find the weakness, to find the hole where I could hit back and stop this from happening. Was I being selfish? Was it only about how much I needed these men in my life? I was confused because, truly, we were all in this together. As someone who loved them, as someone watching out for them, I knew they were unaware of the toll this would take on them over time. They were going to get taken down still deeper, and I refused to abandon them now.

So the verdict was wrong on every level – wrong for me and wrong for the men. They had not yet recovered from the shock, and I knew the worst was yet to come. No one was going to be there for them. And the delicate framing of my emerging self was fragile. I needed more time. We all needed more time. This was insane. At every conceivable level, it was insane.

I was angry that I left myself so open. I struggled desperately to close down, to not feel my feelings around this … around them. I hated these feelings – and my vulnerability.

My friend Dawn tried to slap me out of it. "Jess, you can't go on riding around in fire trucks for the rest of your life. You're not a Dalmatian."

I tried to suck it up and convince myself that it was over, as I knew it would someday be. I would have to go back to my life eventually, but like this … ? Did it have to be done in the cruelest, tear-your-heart-out-of-your-chest way? It was all wrong.

I couldn't sleep. Internally, I felt ripped apart. I sobbed for hours. By 4 a.m. I was exhausted and could no longer fixate; it just hurt too much. In this moment of quiet I finally allowed my mind to go still, and in a split second the warmth of the entity – the "voice" – presented itself.

"Are you done crying yet? Are you done? Because if you are, we have something we want to say to you."

Oh jeez, I was crazy. Imaginings. Talking with air. None of this had been real. For eleven months I deceived myself. I had been making it all up.

Well, I wasn't going to be misled anymore. It was over. My life was reverting back to dead, failure was in sight; sliding down the glass bowl once again. And yet the energy was so insistent and irritating, that I finally gave in.

"Okay, what do you want then?"

There was silence. They lined up in a row and stood beside me. A solid sense of support. It was unmistakably them, and it was unbearable, knowing the door was being closed.

I poured out my heart to the darkness.

"What am I going to do? What am I going to do? I don't want to live my life without you in it."

"Don't worry, you will be with us. It is not yet revealed. Just wait."

I knew I would have to respond to Dan's letter eventually. While he indicated that I was welcome at Engine 32, I wondered what I would do there. Come down and hang out? Like Dawn said, 'be a Dalmation?' Sit there in a corner like Grandma, cracker crumbs around my chair because I hadn't moved all day? I had too much pride for that.

I hoped I did, anyway.

It couldn't be over. It just couldn't. How could the universe lead me through such an experience, allow me to become vulnerable and real for the first time in my life, only to slam the door shut? If being with Engine 32 wasn't right, what was?

I recalled a moment in the kitchen with Manny one evening in June. He was preparing the meal, and I asked him how he had managed to maintain his serenity after all he had experienced. It was he, after all, who had taken us into Ground Zero that night in February. I never forgot the calm on his face that evening, and throughout the entire nine months I had known him.

He replied, "Either God is all or He isn't. If I believe He is, then I have to trust that there is a reason for all of this, and I just can't see it yet."

Manny had showed me in an instant my own hypocrisy around faith. Faith is easy when things are going well, but so hard when they aren't. Dan's letter had shaken me up, carving a deep wound of self-doubt. If I believed in the love I had for these men and the goodness they had brought to my life – if I believed that this extraordinary year was only the beginning of what my life was meant to be – then I needed to have faith. The alternative was to fall, screaming, into the void, perhaps never finding my way back.

My emotions had me seriously unsettled, but I kept them hidden. These men had been through hell, and I didn't want to stress them out with my stupid problems. I knew they would be aware I was hurting and for that reason alone I needed to be as strong for them as they had been for me.

From: Jessica Locke <jessical@shore.net
Date: Tue, 03 Dec 2002 20:59:53 -0400
To: Dan Fullerton <danfullerton@yahoo.com

Dear Dan,

Thank you for your letter. I appreciate the time you must have taken.

I love the Fire Department. I love its purpose, its rules and regulations, its discipline, its customs and traditions, and most of all, its men. I shared with you directly my experiences and my joy of all I learned with each visit.

I was certain of why I came, and what I hoped to accomplish. I gave you all my best.

I understand that you have to get back to "business as usual," although we both know in the shadow of 9/11 nothing will ever be the same again.

As it pertains to me: To disregard the regulations that are part of this Fire Department that I respect so greatly would mean that I learned nothing this past year.

I will not be coming down. I will write the men a letter of closure.

Stay safe. Take care.
Jessica

From: Jessica Locke <jessical@shore.net
Date: Mon, 02 Dec 2002 02:27:23 -0400
To: Tom Laughlin <Tlaughlin@cs.com
Subject: Not coming in December

Tom,

Hey. Dan put an end to it all, but rather than try to interpret his words, I am just going to let you read the letter he sent me. (see attached)

I cried all night, but I'm doing better now. Like I said, to feel safe SOMEWHERE in the middle of this world we're living in – you guys changed my life forever. Thanks for showing me how to laugh again.

If any of you ever need me, I'll be down there in a minute.

Love to you all,
Jessica

12/3/02 4:23 PM, Tlaughlin@cs.com at Tlaughlin@cs.com wrote:

jessica,

wow.

as a sorta new guy trying to do the right thing, a new guy
with four probies to look after and as a some-time-in-the-
future lieutenant, i would have to agree with Dan's assessment.
the new dudes aren't up to snuff (at all) and don't understand
what the other senior members have been through. it can be
fun and games until the doors roll up and the new dudes
don't understand what happens...can happen after. and if my
life is gonna depend on one of them, then they can't be
goofin' off 24 hours at a clip.

also, the chief did come down with a reminder of the rules
and it's been posted for all to know or be reminded of.

sorry. in this case Dan's gotta be the bad guy. but then again,
that's his job to be. perhaps we can charter a weekly bus up to
boston for visits. it'll have to be during the week when buses
are cheaper.

ummm, i dunno what to say. lemme think some more

t

From: Jessica Locke <jessical@shore.net
Date: Tue, 03 Dec 2002 19:14:37 -0400
To: <Tlaughlin@cs.com
Subject: Re: Not coming in December

Tom,

I'm glad to hear that the rules were posted. I can just say to everyone that I have been made aware of this, and want to support this return to "business as usual".

I can't not be sad about this. I'll get over it, though. And anyway, I want to move forward to a bigger level of support for you guys. If I can get this memorial piece off the ground, there will be some royalties to discuss with you how best they can be used for the company. I'd like to see that happen.

I'll keep in touch. You're the best. Thanks for getting back to me so quickly, I feel much better about it all.

Jessica

----
on 12/7/02 3:51 PM, tlaughlin@cs.com at tlaughlin@cs.com wrote:

Jessica

all are aware that you aren't coming down. i'm 50/50 with the whole thing. for the first time to your knowledge i don't have something to say that'll put a smile on your face...or a giggle.

enjoy your holidays away from the tiger den.

tom

----

on 12/8/02 2:35 PM, jessical@shore.net at jessical@shore.net
wrote:

Hi, Tom,

I'm having trouble writing a letter of "closure", I may have
to hold off on it for awhile ... I just don't want certain people
to think I had "abandoned them"--but if I call to talk, I'll
lose it, so ... this is a good thing to get back to discipline and
regulations ... I truly support it.

My love to all of you,

Jessica

----

on 12/9/02 8:25 PM, tlaughlin@cs.com at tlaughlin@cs.com
wrote:

Jessica

where do i start?

in no particular order:

you're truly one of the upper echelon people that e-32

has met in the past 1.5 years. you should know that.

maybe closure is too soon. how about acceptance for the mean time? closure is a lot to ask for. sounds like you're shutting off your feelings. that doesn't really work in the long run. it'll drive you to the devil.

...i'll continue later.

tom

From: Jessica Locke <jessical@shore.net
Date: Thu, 12 Dec 2002 19:17:54 -0400
To: <tlaughlin@cs.com

Tom:

You are an amazing guy. It's an honor to know you. How did you get to be so smart at your age?

Closure, acceptance, neither one fits. But your comments set me straight, I was trying to bury my feelings about it all. Now I'm just numb.

For my money, it's just another loss thrown at the men. It just shouldn't have been handled that way. 9/11 was the most horrific act against humanity since the Holocaust, and you guys were in the center of it. It called for extraordinary measures and extraordinary demands from all of us watching to support you. And it's not time to pull away. This isn't a little scratch that a little band-aid will fix.

As Ray said, this is going to go on for years.

No closure. I will not write that letter.

My love to all of you.
Jessica

----

on 12/13/02 6:30 PM, Tlaughlin@cs.com at
Tlaughlin@cs.com wrote:

jessica,

from this past sun. to wed. we had these 2 german dudes
staying at our house. as monday rolled into tues, and tues
and wed. no one really knew where these guys came from,
who invited them etc...previously we've allowed people into
the house like you or some one representing a town who
made a nice donation. anyway, independent of these dudes,
we had a house meeting last nite and voted no more people
in the house beyond family & girlfriends. that vote had nothing
to do with you and i'm sure the guys who voted no didn't
consider you and what you provided to us. but, more
importantly than Dan's bosses' rules; we voted No on visitors.

it's their loss and they dont even know it.

feel better? ...a little???
now i'm thinking no, you don't.

t

From: Jessica Locke <jessical@shore.net
Date: Sat, 14 Dec 2002 19:17:54 -0400
To: <tlaughlin@cs.com

Tom,

It's okay. You guys have had a rough year. Invaded,
interviewed, photographed, filmed. You could use a
little quiet. I still feel profoundly connected to you all
for some reason, and I don't question it. I'm sure we'll
have reason to see each other further down the line.

You've been great to me, thanks for helping me get
through this transition.

You are an angel.

Love, Jessica

The door to Engine 32 was now closed.
    I was back on my own.

Marge Bell and Stella Jenkins. These two elderly women, members of the church where I sang a solo on Sundays, each gave me an envelope containing a check. Both stated, very explicitly, that I was to use the money *only* for the purchase of a new car. They said I was a good person. They said I was doing a lot of good for a lot of people, and that I deserved their gifts.

The two checks totaled ten thousand dollars.

Ironically, that was just about the amount I had given up due to time off from clients and trips to Engine 32 that past year.

I was completely overwhelmed.

The World Premiere of *Reading of the Names 9/11: The Firefighters* would be performed in a church.

My brother emailed me that the space had little or no room for a choir and so, the fewer singers, the better. I informed him that the piece required a minimum of sixteen singers in order to cover all 343 names. He replied that he didn't have room for sixteen singers; he could barely squeeze in eight.

It occurred to me then that the singers could stand in the outer aisles of the auditorium, surrounding the audience from both sides. That way, the names would be heard in surround-sound.

My brother replied that if I was willing to do that, he would use the entire Kenyon College Choir; fifty-three singers.

Oh my, I thought. This was going to be something.

FIRST PRESBYTERIAN CHURCH
MOUNT VERNON, OHIO
FEBRUARY 8, 2003

I arrived at the church in Mount Vernon just in time for the 1:00 p.m. dress rehearsal. Its interior took my breath away. The focal point was the luxurious silver organ pipes – encased in dark, ornate oak – which spread across the front of the altar. Highly-polished panels enveloped the room and reached ten feet up the walls. The rich, oak pews sloped gently to the steps that led to the dais.

The rest of the church was the purest white, with exposed beams and columns accentuating the height of the room. Elegant, multicolored stained glass windows lined the outer walls, imparting the illusion of flower bouquets. Simple white lights were suspended from the ceiling like glowing stars.

That the venue was a church – and not an auditorium or concert hall – seemed so fitting for the first rendition of this sacred work. It was perfect.

This would be the first time the choir would hear the orchestra, the first time the orchestra would hear the choir, and the first time I would be hearing both. At the last moment, my brother hired a percussionist who had never heard the piece at all.

My sister and her camera were there from Toledo to record the event. We took our seats. The singers lined up in the aisles. The woodwinds, brass, harp and percussion were set up in the choir loft, and the violins, violas, cellos and basses were spread across the dais and floor. The orchestra tuned itself to the oboe's "A", and my brother introduced me to the ensemble.

An incredible excitement builds when music is about to be brought down to the planet for the first time in live performance. Each human, each instrument, the configuration of musicians and singers, and the acoustics of the performance space all combine to create once-only sound and emotional impact. Good or bad, there would be no other performance like this one. The anticipation was unbearable.

The orchestra started and 16 bars in, faltered. They started again. In that moment I was incredibly analytical, looking for everything wrong with my composition while trying to listen to the complex dynamics of the moment. This live performance would allow me the only opportunity to correct any of the flaws my synthesizers could never reveal.

The singing of the names began by the tenors and basses. "Peter Ganci Jr., Orio Palmer, Raymond Downey, Mychael Judge ..."

I made a note to pull the string basses down an octave.

The middle section began with the sopranos and altos. It started as a solo, with one woman singing her list of names. Ironically, for the first time I realized that she symbolized me, coming to the firehouse a year ago to read those first four names: Lt. Robert McMannis, FF. James Gluntz, FF. Elias Keane, FF. Sean Michaels.

The rest of the women joined in, and as they did, the orchestra began to build with them. The men's voices joined the women, and as the names were sung into the air, the room began to swirl like a giant vortex, exerting a powerful force. Wonderfully, amazingly, chillingly we were suddenly being lifted higher and higher, our souls pulled out of our physical beings as though the 343 firefighters had joined together here with us in this moment: Not wanting us to mourn, but attempting instead to rescue us from our grief, to give comfort one more time, to share with us their brotherhood and their unity.

Tears were pouring from my eyes. I looked to my sister to see her tear-streaked face mirroring mine. The choir was crying, the orchestra was crying.

I leaned over and whispered, "I didn't write *this*."

Something extraordinary had occurred. The percussionist approached me after the rehearsal, his face wet. "You know, 9/11 never got to me. I mean, it was a bad thing and all, but I never cried. This piece got to me. I don't know what hit me."

I knew what hit him. Three hundred forty-three firefighters from the Fire Department of New York City.

And this was only the dress rehearsal.

Knox County Symphony Orchestra
Post Office Box 454
Mount Vernon, OH 43050
March 10, 2003

Dear Firefighters of Engine 32:

On February 8, 2003, the 53-voice Kenyon College Chamber Singers and the 60-member Knox County Symphony presented the world premiere of Jessica Locke's *Reading of the Names: The Firefighters* to a packed house. The presentation was unlike any other that I have experienced in my many years as a professional musician. The choir and orchestra performed nobly, even though many said that they lost sight of the music in front of them because of their tears. The audience likewise was affected deeply, so much so that for the first time in my nineteen years with the Symphony the performance was followed by a profound and solemn silence – a silence broken only when I turned and acknowledged Jessica Locke as the composer, resulting in a thunderous standing ovation. Everyone present took notice of the dramatic impact of the composition, but more importantly, we all again felt the enormity of what happened on 9/11. Though (by musical necessity) the name of every single firefighter might not have been consciously understood by the audience, the singers were both diligent in singing each name assigned to them, and in the most reverential manner possible.

I am proud to have been the first person to conduct *Reading of the Names: The Firefighters*. The entire Knox County Symphony Board heartily endorsed this project from the very start. We are happy to follow Jessica's request that her entire fee be donated to you at Engine 32 to be used in any manner you find appropriate.

A music professor at a neighboring college offered the following remarks: "The opening selection by your sister was most moving and such a profound expression! It enabled us all to pay homage vicariously and find a voice to express respect and gratitude and grief." Please know that even as world headlines move on to other topics, people here in Knox County, Ohio, are still aware of the bravery and sacrifice that took place on that day.

Sincerely,

Benjamin Locke, Conductor,
on behalf of the Knox County Symphony Board

*   *   *

A few weeks later my sister sent me the pictures from the performance. In one, everything is very blurry, as though people were being lifted up out of their bodies towards the ceiling. She swears she didn't move the camera.
     I believe her. I was *there*.

BOSTON, MASSACHUSETTS
MARCH 2003

Life went on.

It was an awkward truce. I was stronger for my experiences; certain that the ability to give to and serve others was a more worthy estimate of my value than any college degree, film award or financial net worth. I felt poised and ready to move forward now in any way I chose. The lesson I had learned from Engine 32 was clear: To accomplish anything, I would have to walk into the fear, toward the unknown. Did I want to be a famous singer? I knew how to do it now. Did I want to become a successful film composer? I could do it. I had spent 35 years of my life honing these talents, and nearly 20 years as a bodyworker. It was time to claim my rightful place. I had earned it, and I now had the confidence to do it.

But something was missing. I wasn't happy. And I realized that while there were moments when I achieved some semblance of artistic perfection, there was nobody to come home to and share it with. My cats were not interested if tuna wasn't involved. I had to prove my skills to new film producers again and again. No respect had been built, and even more disheartening, no sense of community.

My thoughts remained with the small firehouse in New York City. Realistically – and financially – there was no way I could move there, nor could I figure out where or how I belonged if I couldn't work in the firehouse. Long ago I learned that true love was about action, not feeling.

Without being able to serve them I was not sure whether this was love, or attachment. I wasn't used to dealing with feelings of want; they always left me uncomfortable. In the past, if I couldn't have something, I shut down. This I could not turn off. Was this reality good or bad?

Resigning myself to what was, I made cookies for the local firehouses and delivered them to the door. "Thank you very much." And again the door closed. I understood. There was no opportunity to prove myself, or to earn their respect. I felt demoted.

The war in Iraq began. I felt the entire soul of the country break down beneath the weight of more death. War begets hate. Hate begets more war. Mankind had learned nothing.

While my vertigo had finally dissipated with the homeopathic remedy, my vision continued to worsen. I made an appointment to see a very gifted vision specialist, Rosemary Gaddum Gordon of Cambridge, Massachusetts. Ten years prior, I had gone to her after suffering from computer-induced eyestrain and had experienced marked improvement.

So again, I sought her help. I apprised her of my extraordinary experiences with Engine 32, and she focused on the feelings of safety they had provided, using it as a foundation to explore that which wasn't safe.

"We lose our vision when there are things we are afraid to see."

As she guided me through exercises which interfered and disrupted my patterned way of seeing, I became more and more uncomfortable emotionally. But I wanted to get to the bottom of my vision problems, and if she could handle the outpouring, I wanted to see whatever it was. She suggested that I not *"try"* to see.

The sexual abuse of children first came to my attention at Ohio State University; a story had been published in the campus newspaper in 1969 about a man who had molested a little girl. It was the first time I had ever read anything about the subject. Back then, people didn't talk openly about such things. I found myself denying that anyone could do that to a child. I felt nauseous, wishing I had never read the article, as uncomfortable and slimy feelings rose up within my own body.

In 1986 the subject resurfaced after the car accident that injured my back and led me to the Alexander Technique. After only three lessons with the teacher, I told him that I was having weird feelings of being sexually abused although I assured him it never happened. Surely, I would remember something like that, I said. I would know it if something like that had been done to me, I said. The fact that I couldn't remember the first five years of my life began to trouble me. Deep inside my psyche, something was crumbling. Thus began a long process of peeling away my outer layers, uncovering a great anger toward my mother, and uncovering myself. I did it all through bodywork.

It was in these revealing years of painful self-examination that I came to realize how trauma can be stored in the muscle and connective tissue of the body where the mind can no longer access it. The Alexander Technique gave me a window through which to observe the possibilities of what I might become, a weekly respite from the horrific stress of my life, and a way of managing and slowly releasing it all.

During that period, I realized that my muscles held memories I didn't yet understand. Throughout the process of release at this cellular level I would consistently find myself locking into a frightening posture, holding my hands up in resistance, my mouth forced open in a rigid "O". My shoulders would press forward while my neck and head were pushed back, as though I was fighting something or someone coming at me. The teacher that I worked with had seen this before. I was open to considering the possibility of sexual abuse, but without a concrete memory, it seemed a convenient excuse for my failures in life. I needed proof. And, even if it did happen, I should have been able to overcome it.

I was in this body posture as Rosemary talked me through the visual re-education. But this time, I *saw* something. Just a flash of something naked over me and the sense of being held down. Something soft, puffy dragged across my chest while I'm pinned down, and I'm pushing away with all my might. My mouth is being forced open, something is being forced into it. I turn my head every which way in avoidance, pulling back with my neck in a violent spasm. My jaw is locked open, and the intrusion enters at an angle. The right side of my head feels like it's going to split open. It resonates as an ache; it is the feverish divide in my head that began on 9/11.

Rosemary asks me to describe what is happening, but I tell her I have no words, I do not have the capacity for words. She uses the term "pre-verbal," and I understand that this is happening when I am too young to speak. The muscle tension is very familiar, a muscular "fix" that I have carried around my entire life. The divisive line within my head that showed up after 9/11 is the sutures of the skull being stretched to their maximum. I surrender to the aggressor; fighting it is hopeless and I am too tired to continue. He does other things to my body, things I am too embarrassed to say aloud. And then it all happens again. This has happened more than once; I have gotten used to it, and know I can't fight it. I can't see who it is. He comes at night. The door opens and with the light from the hallway behind him he exists in shadow, a silhouette. I cannot see his face.

It is not pleasant to bear witness to such a thing, and yet, as I acknowledge the possibility that this actually happened to me, my life starts to make sense. The feelings of shame, of being different, of being terrified of men, of being dirty and ugly and evil had a beginning that had nothing to do with me. I now understand why I may have attracted all of the perverted sexual attention, a sordid reminder of that which needed healing. My mother was not the cause, and yet, as I recall the nights leading up to her death and the terror around her leaving me, I think back to the week she left me with those odd people in the apartment downstairs. She said I simply went berserk after that.

I start to recall other behaviors: Waking up in terror, bracing chairs against the doors of my apartment, staying up until 2 a.m., standing guard. The attacks on 9/11 brought back a 50-year-old lost memory of being violently split in two, physically, emotionally, and spiritually.

*　　*　　*

Two weeks later I see Rosemary again. In the interim I integrate our first meeting. As we begin I find myself going into the defensive rigidity once more. But now I find I am able to go even deeper, further back in time.

In mind's eye I experience myself moments before any of the trauma is perpetuated. There is someone I recognize. It is the glowing woman-child who was reflected in the mirror at Engine 32; a being at peace ... protected, safe, and still innocent.

And then the man comes through the door and he begins, and I watch as he goes at me again. But it doesn't even register now because I know what I have always needed to know:

Once upon a time I had been okay.

The fifty years of pain are gently lifted and gone. No longer do I resonate as ugly and evil. A quiet peace enters my body. I just *am*.

No one can resolve such deep issues until they feel safe enough to go back and face them. If there is one word to describe what the firefighters gave to me, it would be "safety." The men at Engine 32 facilitated my merge into one self, and in doing so, allowed me to force this man out into the open to expose the truth about my past, and free me for my future.

There is a permanent unity within me now. Engine 32 is no longer the glue keeping the two pieces together. They did their job. They rescued me. They gave me back my life. Do they know the extent of the gifts they give, or the thankfulness in the hearts of those they give them to?

MANHATTAN
MARCH 26, 2003

The subway car held a thin crowd of twenty souls this rainy evening. With eyes cast down I rubbed away the tears as fast as they poured out. I hoped that no one would notice my sadness, or my wet, swollen eyes. A few stops later, when the tears had subsided enough for me to see, I looked up. To my amazement, everyone on that car was glancing over at me with tender kindness. Though a year and a half had passed, the soft cores exposed by the tragedy of 9/11 apparently remained, even within notoriously tough New Yorkers.

I had returned to the city to say good-bye. It was time to grow up. My task was to let go of Engine 32, to gain closure on a too-brief time in my life when men were princes and chivalry reigned; a time when a woman was safe and protected. I would always have the memories. I would always know I was the luckiest girl in the world to have had such an experience. If I ever ended up in that nursing home alone, I would have a smile on my face for the rest of my life.

But right now I couldn't stop crying, and I wasn't sure why. Perhaps afraid of what I would find when I went to the firehouse; uncertain of how I would be received. They had voted: *No more visitors except family and girlfriends.* I wasn't family, I wasn't a girlfriend. *Back to business. A tighter ship.* No place for a *massage therapist,* or a woman, or a civilian and *only on the apparatus floor. No visitors after 10 p.m. No more riding in the fire engine.* None of these rules had applied to me. My *office* was on the second floor and now I couldn't even go upstairs. I wasn't sure what would be okay anymore.

I came here to present them with a check; money raised from the performance of *Reading of the Names.* That was a legitimate reason to stop by, right? It was all that saved me from being a Dalmation, although realistically, I could have mailed it. I had no reason to be here other than … I loved them. And that wasn't enough right now, as far as I could tell.

The cold rain dripped from my umbrella as I made my way up Fulton Street. Ground Zero needed to be my first stop in the city. It had been six months since my last visit, and I wanted to check on the lonely emptiness of that place. Was it doing any better? During each of my visits, I always paid my respects. Today I came to reassure the horribly wounded landscape that it, too, would heal with time. It would get better.

As I reached the site, I noticed that the viewing platform was gone, as were the memorials that had grown so wild around St. Paul's church. I stared through the new chain link fence that now surrounded the barren hollow. The harshly-lit concrete landscape before me seemed a monument to 50 years of life without Engine 32. The vicious attacks, the rape, the pain, the isolation, the emptiness: There was no hope here. There was no life.

I had to pull myself together. I could not go to that firehouse all teary-eyed. Not today. Standing in front of Century 21, a store across from Ground Zero, I used my new cell phone to call Beth, an old friend who would understand the complicated emotions that churned within me. She listened with great kindness. The tears abated, and I was ready.

I made my way past dimly-lit puddles, thinking of how much I had changed, how fearless I had become in this world. The path to Engine 32 had been worth dying for. So much good had come to me by daring to walk down these deserted, narrow streets.

I rang the bell. A new probie – such a youthful, innocent face – opened the door. September 11th had not tread upon his heart. At this moment it occurred to me that Engine 32 was going to change; someday all these men would be gone. My heart ached at the thought, that someday I would no longer be known or welcome here. I followed him in, making my way down the narrow path past the fire engine, reaching out to gently caress it as I passed. My beloved Engine 32.

In the dining area was Tommy, Ray, Freddy Murphy, and "Lou", dressed in those intense blue uniforms I loved so much. Freddy, now into his second year on the job and no longer a probie, had changed. A new maturity had emerged, his shoulders had broadened, and his demeanor was more assured. They stepped up to greet me warmly, a hug here, a peck on the cheek there.

"How've you been? Did you have a good trip down? How'd you get here? The train? Amtrak? Where you staying? The condo? Want some coffee?" All the same old questions they asked every time I came down.

I handed them the check. No one made a fuss about it. It was always that way about emotional things, we didn't talk about it. We didn't have to. There was the beautiful old broken-down couch, and the quilts, the stained chairs looking even worse than I remembered and I loved it all. The company journal was on the dining room table with the blue and red pen rubber-banded together. The newspaper article I sent announcing the world premiere was up on the bulletin board, its edges curled up. My brother's letter was alongside.

Covertly, I studied their faces. Though their smiles remained, their faces had changed. They looked so tired, as though slogging through quicksand. They had emerged from shock, and the dead weight of sorrow had finally caught up with them. The "freeze" on their faces and in their eyes was apparent. It killed me, knowing how hard I had tried to prevent this. And I knew I could change it, but I had to sit there politely, like a guest, and do nothing. The words were struggling to emerge. "You need a session. C'mon, upstairs to my office, let's go." I couldn't say it. I was no longer allowed to help.

No visitors above the apparatus floor.

The gathering dissipated, and I was left alone as they resumed their routines – just like the old days. I'm not sure what I expected. After all, they were working, and I couldn't expect them to sit around and entertain me. This is the way it was when I belonged here, when I was their *exclusive massage therapist.* They allowed me to make myself at home; do what I wanted, go where I wanted. No one was showing me to the door. Had I missed my cue? Stayed too long? Should I have said, "I must be going now?" They would know I was lying, that I had no place else to go.

I wasn't sure what to do. I stood up to go to the kitchen, pausing to read the bulletin board. Some of the memos were frightening, alluding to events and circumstances civilians would never be privy to. I shook my head, not knowing how these men were able to withstand the stress. Firefighting was a different job now; terrorism had made certain of that.

I went into the kitchen to get a glass of water, relishing the familiar smells, the warmth of the blue pilot light in the broiler of the ancient stove. I remembered so many pleasurable evenings in this room, the joy of making my first dinner here. I felt a sudden desperation to memorize the details; the fear of losing it forever suddenly becoming very real.

An announcement came over the P.A. They were going out for the meal.

Maybe I should leave now. Watching them drive away without me … I couldn't bear it. I moved to the far corner of the kitchen, feeling terribly exposed, and angry at myself for not having come to terms with the new reality. I still wanted this more than anything else in the world. What was wrong with me?

Freddy came in. "Are you staying for dinner, Jess?"

I was barely able to get the words out. "I am just so grateful to be standing here in this kitchen right now."

Freddy, surprised at the intensity of my reaction, pulled me in for a brotherly hug. I surrendered, realizing I would just have to let myself be here in whatever way they would allow.

We walked together through the dining room and into the garage where everyone was congregating, getting on their bunker gear, stowing equipment into the engine.

And then Tommy looked me straight in the eye and said, "You coming with us?"

The old cue ripped through me; a painful jolt. It was against the rules to take a civilian out in the fire engine. I hesitated.

"C'mon," he said.

He opened the side door and held out his hand.

I saw the determination in his face: Rules or no rules, they were not going to leave me behind. I suddenly understood that the bonds we had forged through an unspeakably painful year were inviolable. It was I who had tried to diminish the strength and depth of our journey together because I couldn't face living without them. Through their wisdom, they let me discover the truth for myself: Their commitment was real. If I didn't know it now, I never would.

I jumped the high step to my place in the back. The engine started up. I felt the husky deep growl of this extraordinary animal as it came alive, the diesel fumes filling the cab, the red lights flashing and reflecting off the walls in glorious celebration. My entire being resonated with the rumbling sounds and pungent smells. I belonged in this fire engine. Whatever it meant, however it would manifest, my life would never again be complete without Engine 32. There was no explaining it – for it made no sense – but I was tired of denying it. The questioning in my soul was relinquished forever, buried by the sound of the red door rising up to the rafters overhead, the men stopping the traffic in the dance I loved, the siren sounding its warning. As we departed the firehouse, I looked at Tommy sitting across from me with that incredible smile on his face.

I had work to do.

FIVE YEARS LATER

JANUARY 13, 2007

There is only one purpose for the writing of this book ... to give back.

The health problems of NYC firefighters who responded to the World Trade Center disaster continue to escalate dramatically since I first stood at the door to Engine 32. The consequences: 75 percent of all firefighters who worked for the FDNY on 9/11 are on the job no longer.

Knowing that my work was still needed, and no longer allowed to work inside the firehouse, I borrowed a massage table and office space and continued to come down to the city as often as possible, considering finances and other obligations. Soon I expanded my program to other firehouses. It was not on any grand scale, but just showing up, still supporting these men, was almost as effective as if I had treated each individually. I began a newsletter, *The FDNY Project*, sent to family, friends and businesses, asking for their support to help me continue my work. I was never disappointed with their generosity and encouragement.

In 2005 the American Red Cross gave a 9/11 Recovery grant to the Olive Leaf Wholeness Center in Manhattan to provide holistic care to first responders. I contacted the Red Cross to see if I could obtain a small grant to assist with the continuation of my work. Because I was not a tax-exempt organization, my request was rejected.

I explained to the Red Cross that I could not refer my men to Olive Leaf if there was any chance they would be turned away; I knew this program would be in high demand once word got out. These firefighters could not have any more doors slammed on them and their needs. I was assured that if Olive Leaf could show the need for more funding, they would receive it.

As of this writing, the waiting list for services at Olive Leaf hovers at 400. The initial funds were exhausted within six months because of high demand, and because so many people were sick. Although participants showed marked improvements in their health and well-being, the Red Cross did not renew Olive Leaf's funding for 2006.

Through the gracious generosity of the law firm of Gotshal, Weil and Manges, I took the steps to become incorporated as the Jessica Locke Firefighters Fund, Inc., a tax exempt 501(c)3 organization. The mission statement allows me to provide Alexander Technique sessions to firefighters. It will address and find ways to eliminate or ease substandard conditions in firehouses, which ultimately add strain to an already stressful occupation. The Fund will work closely with communities to educate and address these conditions, and will solicit donations of funds and/or equipment to alleviate such circumstances.

As the health problems of 9/11 reach crisis proportions, my goal is to use the funds generated by this book to open an alternative health care clinic in New York City where I can continue to oversee the health and financial needs of these firefighters who served above and beyond the call of duty at a time when our country so desperately needed them. We could – and did – rely on them without question, and we will need to again.

It is our turn to take care of them. Not because they would ever ask, but because it is the right thing to do.

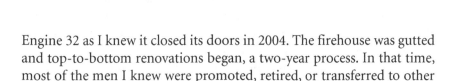

Engine 32 as I knew it closed its doors in 2004. The firehouse was gutted and top-to-bottom renovations began, a two-year process. In that time, most of the men I knew were promoted, retired, or transferred to other firehouses across the city.

Though my Engine 32 is gone, in my mind and heart I have never left. I now understand the meaning of the philosophy lesson at Ohio State. The professor drew a picture of a tree and suggested that because we all agreed in our minds what a tree was, the tree was there. And I said, "There's no money in this, is there."

In the firehouse, everyone agreed in their minds and hearts to hold themselves up to a code of honor. While unwritten, it was nevertheless an invisible touchstone one could sense upon entering the portal. It was a code reflected in how they lived together, worked together, and most importantly, watched out for each other and loved each other. There is no money in philosophy *or* firefighting. You can't pay a man to give his life. But without ideals to strive for, or a code of honor to live by, we are lost as individuals, as families, as communities, as a nation. The unbearably beautiful gift offered in the sacrifice of the 343 firefighters is the path back to ourselves – if we choose to take it.

A FINAL BLESSING

I close this book with tributes from the Memorial Dedication Ceremony for the four men lost from Engine 32: Lt. Robert McMannis, FF. Sean Michaels, FF. James Gluntz, and FF. Elias Keane.

The first is by Captain Dan Fullerton; the second is by the sculptor of the memorial, Robert Girandola.

When it was decided to install a monument to the members that were lost in the World Trade Center, we were encouraged by the desire of people to be involved.
It let us know others knew it was as important as we did.
As time passes and memories fade, it is sometimes a struggle to keep those we have lost alive in our hearts.
This monument will help us to do that.
It is also important to honor those who gave so much of themselves.
This monument is a picture; a picture that tells a story.
It is the story of ordinary men with an extraordinary purpose; to protect life – the very mission of firefighters.
They faced an exhaustive climb, carrying heavy gear and equipment.
Their destination held death and destruction.
Their situation was uncertain – their resolve was not.
They did not hesitate, they did not question.
They met terror with valor, they countered evil with compassion.
They pressed on with dedication and courage.
This monument honors that dedication and courage.
And yet it is not enough.
To truly honor them – to keep them close – we must allow them to inspire us every day.
After you have stood back to admire the beauty of the piece, step up and have a close look, and know what these men knew in that stairway.
Reach up and touch it, and feel what they felt in that tower.
And whenever you have doubts, remember their resolve.
When life's burdens are heaviest and you grow weary, remember their dedication.
And if ever you are afraid, remember most their courage.

– Captain Dan Fullerton

A person is what they do – no more, no less
    A person can become what they admire.
I was at work on that day
I heard the towers had collapsed
    I ran home – I was 80 miles away from the danger
    And I ran home
I put on the television – I watched as my head swam
Fear, confusion, anger, frustration, sadness
I waited impatiently for my family to come home
And when they did
    We hugged
    We cried
    We watched the television and waited to see if other, braver
men and women would make things right in the world again.
We are here to honor four such men –

    Lt. Robert McMannis
    FF. James Gluntz
    FF. Elias Keane
    FF. Sean Michaels

All of the 343 active duty firefighters to have lost their lives that day
And all of the brave men and women who put fear aside and chose to
make things right again.
I deeply admire you.

In order to create this piece – I listened to the stories of the men of
Engine 32
I read the letters their loved ones wrote

I watched that day unfold again and again
and, amidst the cloud of pulverized concrete and glass
an image emerged.

I saw fallen heroes and the walking wounded –
Men and women who spent their days 'hoping and praying, digging,
searching, thinking, sweating, tearing and more hoping praying.'
I saw lives ruined and families rocked and whole communities racked
with sorrow.
And I saw men and women who chose to do something about it.
I asked Matt Dunn, the only man from Engine 32 to make it out of the
North Tower, what were you most proud of that day?
He said – 'We stood fast'
In the face of overwhelming destruction, when a nation sat in dumb-
struck fear these men stood fast – and then, proceeded up the stairs to
'put out that fire.'
Over thirty flights of stairs,
Turn out coats,
Soaring temperatures,
Tools, roll-ups
I asked Matt – 'How did you guys go on?'
He said, 'Well, we kind of drew strength from each other.'

These men may not have been perfect, but they became perfect.
At that moment and in that hour
And it was this STRENGTH that transcended the darkness and the evil
that was transpiring above them.

I have read of firefighters despairing that they themselves did not save
enough or very little –
To you MY brothers, I must say – your mere presence brought comfort
to those in the last moments of their lives.

You INSPIRE in the minds and hearts of those around you an idea –
the idea that you may not be perfect – but you will try to be –
That you may not be able to save – but you will try to save –
For in the end ALL of the men and women who choose to save –
And these four men –

      Lt. Robert McMannis
      FF. James Gluntz
      FF. Elias Keane
      FF. Sean Michaels

you bring the PROMISE of hope,
You DRAW STRENGTH from each other –
May we draw on that strength so that we might have the courage to do
what we should do
in the hour of our calling.

– Robert Girandola

ACKNOWLEDGEMENTS

Until you write a book, you never understand why writers thank their editors first. The editor keeps you honest, makes you look good, stands up to you when necessary, acquiesces when you are certain of your words. In all these ways, I will never be able to thank Maria Scarfone enough for her artistic vision, care, and reverence in the creation of this manuscript.

To Sue Kilrain whose generosity touched so many; without you none of this could have ever happened. Thank you so much.

To Dawn Carroll; your character shines like a light throughout this book, as well as within the many years I've been blessed by your friendship. Shine on.

To the unsung heroes of the *FDNY Project*: Your generous gifts and support cannot be measured. When thoughts and prayers were not enough, you came through for the firefighters at the time when they needed it most: Andrew Allner, Luis Barragan, Brooke Barton, Marge Bell, Amar G. Bose, Fabrizio Bottero, Crandall Bowles, Ross Bradford, Marshall Brown, Fred Carl Jr., Barry Cohen, Dean Crabtree, Jen Davis, Tracy Davis, Judith Diamond, John Drillot, Tom Dunn, Ann and Bob Everett, Rhya Fisher, Frank Fleischman, Maria De Jony Franz, Barbara Fraunfelder, Peter Fromme-Douglas, Stephen Gagnon, Herb Gellis, Beth Ginga, Leon Gruenbaum, Barry Haber, Anne Hammel, David Hertz, Walter Hertz, Stella Jenkins, Bea Jouett, Kristine Juster, Patrick Kane, Ian and Rebecca Kasarjian, Kenyon College Choir, Andrew Kelly, Kevin Kertscher, Dena Knop, Knox County Symphony Orchestra, Laurie Levine, Lois Lewis, Peter Lieberman, Benjamin and Kay Locke, Justin Locke, Rebecca Locke-Gagnon, Anna White Lovely, Judy Lugo, Christel Mann, Alison McLea, Ann McLea, Katherine McGuire, David McIlquham, Bill McLaughlin, Maria Ng, Nick Ord, James Orent, Lisa Pacitto, Iris Pinlac, James Polark, Jean Putnam, Brian Radovich, Kelly Rampson, Elsa and Chuck Reeves,

Stephen Ruggere, Michael and Anu Sheridan, Bonnie Staffel, Steve Simon, Shawn Slattery, Jeanine Spina, Gloria Spitalny, David Suh, Lester Tatum, Steve Terry, Tom Tervo, Rose Thomas, Glen Tuomaala, Donna Whipfley and the members of the Masterworks Chorale, Elliot Williams, David Wolf.

My deepest appreciation to Charlie Eitel and Simmons Bedding Company, who gave above and beyond and brought great comfort to the FDNY firefighters.

To Andrew Troop, Shayla K. Hargrev, Joseph H. Newberg, and Jill D. Chesler, attorneys at Weil, Gotshal & Manges; Thank you for believing in my mission, and for your continued support and advice.

Special thanks to my teachers, who showed me that music, healing, truth and love were all the same language: Laurie Nelkin, Arnold Siegel, Elsa Reeves (my new "mom"), Tommy Thompson, Robert Lada, Debi Adams, Robert Gerle, Betsy Politan, Lennie Peterson, Richard D. Lang, Daniel Pinkham, Joseph Gifford, Rosemary Gaddum Gordon, Burdette Green.

To Bill Gorsica; your kindness will never be forgotten.

To Shannon, Stacey, and Mike: you probably learned more about the fire department than you ever planned, thanks for listening.

My grateful appreciation to the Newton, MA Fire Department for helping me expand my role as advocate. I love you all.

To Drew Allison; thank you so much for the inspiring and beautiful book design and website, and a twenty year friendship that only gets better as we grow older.

My heartfelt thanks to John H.M. Austin, M.D. and Pearl Ausubel for allowing me to quote from your research on the Alexander Technique. Acknowledgement of this work is long overdue.

Craig Williams of the New York State Museum at Albany; many thanks for allowing me the privilege of access to the archives.

To Alicia E. Vasquez for the extraordinary poem *Don't Look For Me Anymore;* thank you. Your literary paintings humble me.

Rob and Tina Girandola; I'm honored to be walking the planet with you. Your family gives me hope for America.

To Lt. Joseph McMahon; thank you for the beautiful synopsis on the back cover and your overwhelming support of this project.

With great appreciation to the following members of the FDNY for their support, guidance and research: James D., Richie R., Robert M., Kevin H., Nick K., Scott F., Rich N., and especially Jay J. and Dan A.

To Arlene B.; you are a being filled with love and grace and extraordinary strength. I am so blessed to have met you. I will never be able to thank you enough for your kindness and generosity.

Officer Weber, wherever you are, I'll never forget the calming and caring sound of your voice against that terrible landscape. Thank you for sending me down the right path. I hope to meet up with you again someday.

To Tommy "Robin" Laughlin, who had this crazy idea that I could actually write a book; your belief in me gave me the courage to "keep on a-chuggin." I still believe I can do anything because of you.

To Matt and Elliot; I am so grateful to be able to include your incredible stories within this manuscript. Thank you.

To Dan Fullerton; there is the rank of "Captain" and then there is a quality of leadership which can only come from the heart. You had both. Thank you so much for allowing me to use your emails, and for all your help and research concerning this book.

To all the members of the FDNY who invited me into the warmth of their firehouses, shared their meals and told me their stories; I will never forget you.

And lastly, to the members of my beloved Engine 32: Dan, John, Ray, Elliot, Matt, Steve, Mark, Richie, Manny, Joel, Lennie, Ronnie, Harry, Tommy, Scott, Freddy, the many Lou's, and the others who were not mentioned in this book: You taught me the meaning of family, respect, commitment, and love. Thank you for trusting me "to do the right thing" with this manuscript.

You truly *are* angels.

Forever in my heart, you guys are the BEST.

God Bless.

"I just wanted to thank you from the bottom of our hearts for all you have done for us, and all the other companies you have helped out. You are the one person who has really stuck with us all this time. We could never thank you enough for all the help you have given us. But please know this, you did make me feel so much better."

– NYC Firefighter

"You have given me the help that I needed. And, maybe even more important, you've given me the info/resources so I can help myself through testing/diagnosis treatment and healing. I now feel as though I can begin to move forward with some sense of direction & power over my situation. There are no words to thank you."

– NJ Firefighter

"Your dedication to us is greatly appreciated and is always remembered. Thank you for keeping us in your heart."

– NYC Firefighter

Dear Reader,

With the dawn of 2010, the firefighters who worked at the World Trade Center on 9/11, and at the site in the months after, continue to suffer with chronic lung disease, a host of cancers, post traumatic stress disorder, and other debilitating illnesses. At the same time, programs to assist with medical and financial resources are being discontinued.

It is difficult for these rescuers to ask for the help they desperately need, and they shouldn't have to. The mission of The Firefighters Fund is to ensure that these brave men and women are not overlooked, and never forgotten.

Won't you send your tax-deductible donation today to help provide direct support, referral services, and rehabilitation to firefighters in their time of need?

Donations can be made at: www.firefightersfund.org.

Checks can be mailed to:

Jessica Locke Firefighters Fund
505 Waltham Street, Suite 2
West Newton, MA 02465

Jessica Locke is available as a public speaker for your organization or group.

Email: jessica@firefightersfund.org or call 617-924-9999

to hear a recording of
Reading of the Names 9/11: The Firefighters:
please go to www.firefightersfund.org

for information on the Alexander Technique:

Alexander Technique International
1692 Massachusetts Ave, 3rd floor
Cambridge, MA 02138, USA
www.ati-net.com
ati-usa@ati-net.com

for information on Captain Patrick Brown:

www.missyoupat.org

recommended books and videos:
Miss You, Pat by Sharon Watts
Closure by Lieutenant William Keegan, Jr.
Report from Ground Zero by Dennis Smith
Firehouse by David Halberstam
Collateral Damages DVD Etienne Sauret